Studies in Computational Intelligence

Volume 1140

Series Editor

Janusz Kacprzyk, Polish Academy of Sciences, Warsaw, Poland

The series "Studies in Computational Intelligence" (SCI) publishes new developments and advances in the various areas of computational intelligence—quickly and with a high quality. The intent is to cover the theory, applications, and design methods of computational intelligence, as embedded in the fields of engineering, computer science, physics and life sciences, as well as the methodologies behind them. The series contains monographs, lecture notes and edited volumes in computational intelligence spanning the areas of neural networks, connectionist systems, genetic algorithms, evolutionary computation, artificial intelligence, cellular automata, self-organizing systems, soft computing, fuzzy systems, and hybrid intelligent systems. Of particular value to both the contributors and the readership are the short publication timeframe and the world-wide distribution, which enable both wide and rapid dissemination of research output.

Indexed by SCOPUS, DBLP, WTI Frankfurt eG, zbMATH, SCImago.

All books published in the series are submitted for consideration in Web of Science.

Grażyna Szkatuła · Maciej Krawczak

Bidirectional Comparison of Nominal Sets

Asymmetry of Proximity

Springer

Grażyna Szkatuła
Systems Research Institute
Polish Academy of Sciences
Warsaw, Poland

Maciej Krawczak
Systems Research Institute
Polish Academy of Sciences
Warsaw, Poland

WIT Academy
Warsaw, Poland

ISSN 1860-949X ISSN 1860-9503 (electronic)
Studies in Computational Intelligence
ISBN 978-3-031-53098-2 ISBN 978-3-031-53096-8 (eBook)
https://doi.org/10.1007/978-3-031-53096-8

This Springer imprint is published by the registered company Springer Nature Switzerland AG
The registered company address is: Gewerbestrasse 11, 6330 Cham, Switzerland

Paper in this product is recyclable.

Preface

The analyses presented in this volume suggest that the assumption of symmetric proximity of objects should not be established in advance. The assumption of asymmetric proximity seems to be more appropriate for solving many real problems, arising, for instance, in psychology, especially in relation to modeling human judgments of similarity. Directional proximity leads to more effective procedures which can be used when comparing the sets of objects.

As objects we mean, for example, concepts, patterns, references, etc., and description of such an object is represented by a respective set of nominal-valued attributes. However, the values of an attribute may not be exactly known or may repeat themselves. We may have some extra information, for example, what attribute value is possible or acceptable, as well as some information, what attribute value is impossible or unacceptable. Considering the different types of sets allows us to model such imperfect as well as incomplete and inconsistent data. Thus, the nominal values of an attribute may be represented through different kinds of set's elements.

Determining a good measure of proximity between the compared objects is very important both for practical tasks and for theoretical analyses. As a result of comparison, the proximity of objects is understood either as similarity of these objects or distance between them.

The task of comparing such objects can be transformed into a task of comparing the sets of attribute values that describe the objects considered. In the book, we gathered known in the literature, representative proximity measures, which give symmetrical or asymmetrical results for

- the ordinary sets (i.e., sets in which the elements are unitary and belong or do not belong, for sure, to the set under consideration),
- the multisets (i.e., sets in which multiple occurrences of the same element are allowed),

- the fuzzy sets (i.e., sets in which degrees of membership of elements are determined), and
- the intuitionistic fuzzy sets (i.e., sets in which both degrees of membership as well as non-membership of elements are determined).

The elements of these sets are of qualitative nature.

At the same time, we described our original concept of the perturbation of one set on another set, and the measures of directional set perturbation were compared to proximity measures known in the literature. The concept of set perturbation is understood as impact of one set on another set. In other words, the perturbation arises upon attaching the first set to the second set and checking how much such attachment has affected the second set (i.e., how much the second set was perturbed by attaching the first set). Of course, the counterpart case also must be considered, that is - the effect of adding the second set to the first (i.e., disturbing the first set by adding the second set).

The proposed perturbation methodology and the directional proximity related to it include the possibility of designing a relatively simple and intuitive method of matching objects. The measures of directional object perturbation are introduced and discussed. The proposed measure allows us to formulate some additional conclusions relating to the proximity of objects under consideration, due to its asymmetric nature, in general.

The content of the book is as follows. Notations used in the book are introduced and the general remarks about object proximity are included in Chap. 1.

There are several important issues, where direction of comparisons of multi-attribute qualitative objects cannot be neglected (e.g., in psychological literature, especially related to modeling of human similarity judgments), thus the phenomenon of the asymmetric nature of object proximity is discussed in Chap. 2.

The sets of nominal values of an attribute can be represented in different ways, considered in consecutive chapters of the book.

In Chap. 3, the sets of attribute values represented by ordinary sets are considered, and in Chap. 4 represented by the sets with binary coding. In Chap. 5, the sets of attribute values represented by multisets are considered. In Chap. 6, the sets of attribute values represented by fuzzy sets are considered. In Chap. 7, the sets of attribute values represented by intuitionistic fuzzy sets are considered.

Much of the book is devoted to definitions, corollaries, and mathematical proofs describing many important properties of the set perturbations. The text is supplemented with many examples to make it easier to follow, and a rich literature of the considered subject. Additionally, significant parts of the book can be treated as brief introduction to the considered kinds of sets.

The book is intended for data science professionals, philosophers as well as cognitive psychologists, who struggle with practical problems in which asymmetry of proximity of objects cannot be neglected. Thus, we are equipped with a tool to determine

"How much one object is similar to another object",

not just whether

"How much similar are two objects".

The use of the proposed measures of perturbation between compared objects can be very important in data mining or in exploration of Internet resources. Also, we are often interested in distinction as to which object is the prototype and which objects are only its variants.

The book is aimed, as well, at data science specialists and graduate students, who would like to broaden their perspective on the phenomenon of the asymmetric nature of objects proximity.

Warsaw, Poland
December 2023

Grażyna Szkatuła
Maciej Krawczak

Contents

List of Figures

List of Tables

Chapter 1
Introduction and Main Assumptions

1.1 Notation

First, we provide the main notations used in the book:

i	Index of set (value, attribute, object); j, l and n are also used,
A, B, C	General designation of sets (of objects),
U	Finite set of objects,
$card(.)$	Cardinality of a set,
$\prec_L$	Sharp lexicographical ordering of ordered pairs of numbers,
$\preceq_L$	Lexicographical ordering of ordered pairs of numbers,
$Agg(., ..., .)$	Aggregation operation that assigns a real number to any K-tuple of real numbers, $Agg : [0, 1]^K \rightarrow [0, 1]$,
V	Finite set of nominal values $\{v_1, v_2, ..., v_L\}$, where L is total number of elements of the set,
S	Total number of subsets of the set V,
$\mu_A(v)$	Membership function of element $v \in V$ with respect to the set A, that is $\mu_A : V \rightarrow [0, 1]$,
$\nu_A(v)$	Non-membership function of element $v \in V$ with respect to the set A, that is $\nu_A : V \rightarrow [0, 1]$,
$\pi_A(v)$	Indeterminacy of information about element $v \in V$ membership in the set A, $\pi_A(v) = 1 - \mu_A(v) - \nu_A(v)$,
$D(., .)$	Distance (dissimilarity) function between a pair of sets,
$S(., .)$	Similarity function between a pair of sets,
$\hat{S}(., .)$	Extended similarity function between a pair of sets,

G. Szkatuła and M. Krawczak, *Bidirectional Comparison of Nominal Sets*,
Studies in Computational Intelligence 1140,
https://doi.org/10.1007/978-3-031-53096-8_1

$\{a_1, a_2, ..., a_K\}$	Set of nominal-valued attributes a_j, $j = 1, 2, ...K$, which describe objects,
V_{a_j}	Domain of attribute a_j, $V_{a_j} = \{v_{1,j}, v_{2,j}, ..., v_{L_j,j}\}$, where L_j denotes the number of nominal values of attribute a_j,
e_n	nth object, $e_n \in U$ for $n \in \{1, 2, ..., N\}$, where N is the total number of objects,
G_{e_n}	Description of the nth object e_n, represented by the ordered collection of sets,
$G(U)$	A family of subsets of U,
g_i	ith group of objects belonging to the set of objects U, that is $g_i \subseteq U$ for $i \in \{1, 2, ..., card(G(U)\}$, where $card(G(U))$ is the total number of groups of objects considered,
G_{g_i}	Description of ith group of objects g_i, represented by ordered collection of sets,
$t(j, e_n)$	Index specifies, that attribute a_j, $j \in \{1, 2, ..., K\}$, takes value $v_{j,t(j,e_n)} \in V_{a_j}$ for nth object e_n, $n \in \{1, 2, ..., N\}$,
A_i	ith ordinary set, $A_i \subseteq V$ for $i \in \{1, 2, ..., S\}$; A_j is also used as ordinary set,
A_i^C	Complement of the ith ordinary set A_i; A_j^C is also used,
$(A_i \mapsto A_j)$	Perturbation of the jth ordinary set A_j by the ith ordinary set A_i,
$Per^1_{OS}(A_i \mapsto A_j)$	Measure of perturbation type 1 of jth ordinary set A_j by ith ordinary set A_i,
$Per^2_{OS}(A_i \mapsto A_j)$	Measure of perturbation type 2 of jth ordinary set A_j by ith ordinary set A_i,
$[w^i_1, w^i_2, ..., w^i_L]$	ith L-dimensional binary vector,
$V^m(L)$	Set of multisets drawn from the set V, where L is the cardinality of the set V and m specifies maximum of occurences of elements,
S_i	ith multiset, $S_i \in V^m(L)$; S_j is also used,
$k_S(.)$	Counting function (multiplicity function), $k_S(v)$ specifies the number of occurrences of element $v \in V$ in multiset S,
$(S_i \mapsto S_j)$	Perturbation of jth multiset S_j by ith multiset S_i,
$Per^1_{MS}(S_i \mapsto S_j)$	Measure of perturbation type 1 of jth multiset S_j by ith multiset S_i,
$Per^2_{MS}(S_i \mapsto S_j)$	Measure of perturbation type 2 of jth multiset S_j by ith multiset S_i,
$FS(V)$	Collection of all fuzzy sets in a crisp set V,
B_i	ith fuzzy set $B_i \in FS(V)$; B_j is also used as fuzzy set,
$(B_i \mapsto B_j)$	Perturbation of jth fuzzy set B_j by ith fuzzy set B_i,
$Per^1_{FS}(B_i \mapsto B_j)$	Measure of perturbation type 1 of jth fuzzy set B_j by ith fuzzy set B_i,
$IFS(V)$	Collection of all intuitionistic fuzzy sets in a crisp set V,
C_i	ith intuitionistic fuzzy set $C_i \in IFS(V)$, C_j is also used as intuitionistic fuzzy set,
$(C_i \mapsto C_j)$	Perturbation of jth intuitionistic fuzzy set C_j by ith intuitionistic fuzzy set C_i,
$Per^1_{IFS}(C_i \mapsto C_j)$	Measure of perturbation type 1 of jth intuitionistic fuzzy set C_j by ith intuitionistic fuzzy set C_i,
D_{Per}	Matrix of measures of perturbations, $D_{Per} =[d_{ij}]$, $i, j \in \{1, 2, ..., N\}$, where N is number of sets considered.

1.2 Remarks on Object Proximity

A comparison of two objects may be viewed as an attempt to determine the degree to which they are similar or different, according to some predetermined sense. Defining a good measure of proximity or similarity, or dissimilarity between objects, described by values of attributes, is very important both in practical tasks as well as in theoretical considerations related to all kinds of data analysis. Each object is usually represented as a point in Cartesian coordinates, and therefore the distance between points reflects the similarity between respective objects. In general, the space is assumed to be Euclidean, and a distance assigns a nonnegative number, But at the same time, from another point of view the object is described by sets of features.

The term *similarity* is perhaps most frequently used as denoting a compatibility measure of objects, but, definitely, it is both universal and difficult to define. The analysis of the similarity of two objects plays a fundamental role in theories of knowledge and behavior, machine learning models, and human perception. For instance, Attneave in 1950 stated that [1]

"It is difficult to pursue any path of psychological inquiry without encountering the problem [of similarity] in one guise or another",

and

"It is obvious that when things are similar,
they are similar with respect to something".

Evaluation of the proximity of compared objects remains a standing problem in many areas. The role of similarity or dissimilarity of two objects is fundamental in many theories of cognitive as well as behavioral knowledge, and therefore for comparison of objects different measures of object similarity are commonly used. However, in a famous critique, Goodman [2] dismissed similarity as scientifically useless notion. He claimed that the concept of similarity of one object to another object is ill-defined, because it does not include the concept

"under what terms".

It seems to be obvious that objects are similar

"with respect to something".

Defining a sound similarity measure is important in practical application tasks like clustering ecologically related species, in biology, ethnology, taxonomy biometrics an so on (see e.g., [3]).

Cognitive visual illusions are very good examples of complexity of perception of reality (e.g., [4]). *A contrast effect* either strengthens or weakens our perception, for example, *a simultaneous contrast effect* depends on how different colors influence each other.

Simultaneous contrast of colors was described in the 19th century by Chevreul in his famous book on color theory, *The Principle of Harmony and Contrast of Colors*, published in 1839 (translated into English in 1854). However, simultaneous contrast effects were familiar to painters since the Renaissance.

Fig. 1.1 Simultaneous contrast effect of colors

Figure 1.1 shows two identical images, but shown against different backgrounds. Each image consists of a gray house with a red roof and a brown chimney, and the Moon. It must be emphasized that sizes, colors and shades of each image are exactly the same, while colors of the backgrounds are different. The left hand side background is gray, while the right hand side background is black. It is easy to notice the simultaneous contrast effect, namely the right image of the house seems to be significantly brighter and larger, and the Moon seems to be brighter compared to the left image. Since we seldom can see colors separately, our perception of colors and shadows may substantially depend on the background color, and our perceptions do not always correspond to reality.

In most models of similarity it is assumed that the relation of similarity is symmetric. For example, Shepard [5] considered spatial objects and assumed that distance between two objects is equal whatever direction is applied. For years, the geometric interpretation of similarity was in common consideration, where corresponding space is metric (e.g., Carnat [15]; Shepard [6]; Torgerson [7, 8]), in which objects are treated as points.

It is very common to assume that similarity between two objects is assigned as a value from the interval [0, 1], and dissimilarity is defined as non-similarity (i.e., the difference between 1 and the similarity). However, some research indicates that humans judge similarity and dissimilarity in different ways, and people often think that non-similarity and dissimilarity are not synonymous. In the case of considering similarity the common features are emphasized, whereas in the case of considering dissimilarity the distinctive features are assessed.

From the mathematical point of view, distance is defined as a quantitative degree of how far apart two objects are, and synonyms of distance include dissimilarity. Dissimilarity expresses the degree in which two objects are found to be dissimilar, the degree usually ranging between 0 and 1. Each object can be represented as a point in coordinate space, and the distance between two of such points reflects dissimilarities between the respective objects. For metric space, a distance is a function $\mu()$ which assigns to every pair of objects A and B a nonnegative number, which is called their distance, and satisfies the following axioms:

- $non - negativity\ property$: $\mu(A, B) \geq \mu(A, A) = 0$;
- $symmetry$: $\mu(A, B) = \mu(B, A)$;
- $triangle\ inequality$: $\mu(A, B) + \mu(B, C) \geq \mu(A, C)$.

Much work has been done to determine distances of objects described by continuous-valued attributes. However, in general, handling proximity of objects described by nominal-valued attributes is considered to be much more difficult. Namely, nominal values have neither a natural ordering nor an inherent order, and are measured on nominal scales. It is said that an attribute is nominal if it can take one of a finite number of possible values and, unlike ordinal attributes, the set of these values has not internal structure. The case where a nominal attribute can only take one of the two possible values, is usually called binary or dichotomous.

For instance, let us consider the attribute $taste$, which may take values like $salty$, $sweet$, $sour$, $bitter$ or $tasteless$. In the case when the attributes are nominal, definitions of similarity (or dissimilarity) measures become not trivial at all, and for nominal-valued attributes the task of comparing one object with another can be considered in the form of whether the objects have the same or different nominal values. For comparing of this type of data there are two main approaches, namely:

- $Simple\ matching$ - for two possible values (i.e., each value represents different compared object) of the same attribute the dissimilarity is defined as zero if they are identical, and one otherwise. Upon repeating this procedure for all attributes, the ratio of the number of matched elements and the total number of attributes is calculated.
- $Binary\ encoding$ - creating binary-valued attributes instead of nominal attributes, and then any conventional matching method (e.g., the simple matching coefficient or Jaccard's coefficient) can be employed. This methodology of transformed binary attributes does not preserve the semantics of the original attribute and, additionally, the dimensionality of new attributes may increase significantly.

The interest in comparing the collections of objects was already reported at the very beginning of the 20th century. One of the oldest and common known measures of similarity of objects is $the\ Jaccard\ coefficient$, originally called "coefficient de communauté" (coefficient of commonality) by Paul Jaccard in 1901 [9]. This coefficient (now called the Jaccard coefficient, or the Jaccard index) is a measure of similarity and it has been used extensively, largely due to its simplicity and intuitiveness (e.g., Shi [10]; Magurran [11]). Many measures are based on the Jaccard coefficient, for example, the Sorenson index in botany, also known as the Dice coefficient, applied in ecology, and the Czekanowski measure used in his system of racial classification, the F1 score as measure of a test's accuracy, the coincidence index used in cryptography, and others, like the Ochiai and Kulczynski measures. Calculation of these indices is relatively simple and intuitive, and often provides useful results (cf. Wolda [12]; Hubálek [13]). Hubálek [13] lists the Ochiai and Kulczynski indices as providing good results, but the Jaccard or Sorenson measures are typically more recommended and they are more commonly used.

There is a known approach for defining a distance of nominal attributes named the Value Difference Metric (VDM). The approach was introduced by Stanfill and Waltz [14] to provide an appropriate distance function for nominal attributes, and is valid only for the problem of learning from examples (e.g., [16]). The VDM distance between two nominal values is the difference between the estimated decision probabilities, corresponding to these values. The estimation of the decision probability is done on the basis of the training set. For example, we want to distinguish apples among other fruits. The used attribute is *color* with three values: *red*, *green* and *blue*. The pair *red* and *green* would be considered as closer than a pair *red* and *blue* because the former pair has similar correlations with apple classes. One of the main problems of the approach arises if *a strange pair* of attribute values never appears in testing sets. It is worth to notice that the VDM approach is not a metric (as the measure is not symmetric).

Here, we consider a pair of objects A and B, considering a distance measure and the similarity between them. Generally, a distance represents a quantitative degree showing how far apart two objects are. At the same time, similarity describes the degree indicating how close the objects are. It is important to notice that similarities focus on matching relations between non identical objects while the differences focus on the mismatching of objects.

On the basis of the set theory, the observed similarity of two sets A and B can be expressed as some real valued function of three arguments: the set of features common for both sets, the features of the first set that are not shared by the second set and the features of the second set that are not shared by the first set. Using these three arguments we can measure the degree to which two objects (viewed as sets of feature values) match each other.

In the majority of theoretical works on object similarity there is an additional assumption of symmetry of object proximity (i.e., the proximity of the object A to the object B is equal to the proximity of the object B to the object A). The assumption of symmetry underlies essentially the majority of theoretical treatments of similarity.

Some investigations, however, do not accept this assumption, for instance in psychological literature there are two main approaches: the distance model of similarity and the content model. The issue of symmetry was extensively analyzed by Amos Tversky [17]. He considered objects represented by sets of features (attributes), and proposed two *contrast models* wherein similarity was described as comparison of features. Tversky provides empirical evidence of asymmetric similarities and argues that similarity should not be treated as a symmetric relation. There is no uniform concept of similarity that is applicable to all different experimental procedures used in the comparison of objects. So, his models do not define a single similarity scale, but rather a family of scales characterized by different values of parameters.

In fact, quite a number of approaches to modeling asymmetric proximities have been proposed in the literature. They can be divided into three main groups.

- In the first group, researchers perform some preprocessing of data to obtain symmetry. According to Beals et al. [18],

> "if asymmetries arise, they must be removed by averaging or by an appropriate theoretical analysis that extracts a symmetric dissimilarity index".

- The second group of researchers claim that ignoring the asymmetry discards the potentially valuable information. They explicitly model the asymmetries, for example, Zielman and Heiser in the paper [19] decompose asymmetric proximities into a symmetric and an asymmetric component.
- In the third category of approaches, the asymmetries are represented as the directional distances. In such a point of view, asymmetric proximity data are treated in accordance with the original form of the data, and analyzed in view of the asymmetry (e.g., like in the papers of Krawczak and Szkatuła [20–22]).

In the present book, we consider a finite, non-empty set of objects, and each object is described by a set of nominal-valued attributes. The task of comparison of such objects can be converted into a task of comparison of sets of attribute values that describe objects. Additionally, following Tversky's suggestions about possible asymmetric nature of similarities between objects [17, 23] we want just to verify symmetry of objects' proximity using the perturbation methodology.

References

1. F. Attneave, Dimensions of similarity. Am. J. Psychol. **63**(4), 516–556 (1950)
2. N. Goodman, Seven strictures on similarity, in *Problems and Projects* (Bobs-Merril, New York, 1972), pp. 437–450
3. S. Choi, S. Cha, C.C. Tappert, A survey of binary similarity and distance measures. Syst. Cybern. Inf. **8**(1), 43–48 (2010)
4. M. Tanca, S. Grossberg, B. Pinna, Probing perceptual antinomies with the watercolor illusion and explaining how the brain resolves them (PDF). Seeing and Perceiving **23**, 295–333 (2010)
5. R.N. Shepard, Stimulus and response generalization: a stochastic model relating generalization to distance in psychological space. Psychometrika **22**, 325–345 (1957)
6. R.N. Shepard, Stimulus and response generalization: tests of a model relating generalization to distance in psychological space. J. Exp. Psychol. **6**, 509–523 (1958)
7. W.S. Torgerson, *Theory and Methods of Scaling* (Wiley, New York, 1958)
8. W.S. Torgerson, Multidimensional scaling of similarity. Psychometrika **30**, 379–393 (1965)
9. P. Jaccard, Distribution de la flore alpine dans le bassin des Dranses et dans quelques régions voisines. Bulletin de la Société Vaudoise des Sciences Naturelles **37**(140), 241–272 (1901)
10. G.R. Shi, Multivariate data analysis in palaeoecology and palaeobiogeography-a review. Palaeogeog. Palaeoclimatol. Palaeoecol. **105**, 199–234 (1993)
11. A.E. Magurran, *Measuring Biological Diversity* (Blackwell, Oxford, 2004)
12. H. Wolda, Similarity indices, sample size and diversity. Oceologia **50**, 296–302 (1981)
13. Z. Hubálek, Coefficients of association and similarity, based on binary (presence-absence) data: an evaluation. Biol. Rev. **57**, 669–689 (1982)

14. C. Stanfill, D. Waltz, Toward memory-based reasoning. Commun. ACM **29**, 1213–1228 (1986)
15. R. Carnap, *The Logical Structure of the World* (University of California Press, Berkeley, 1928/1967)
16. A. Wojna, Analogy-based reasoning in classifier construction.,in *Transactions on Rough Sets IV*, ed. by J.F. Peters, A. Skowron. Lecture Notes in Computer Science, vol. 3700 (2005), pp. 277–374
17. A. Tversky, Features of similarity. Psychol. Rev. **84**(4), 327–352 (1977)
18. R. Beals, D.H. Krantz, A. Tversky, The foundations of multidimensional scaling. Psychol. Rev. **75**, 127–142 (1968)
19. B. Zielman, W.J. Heiser, Spatial representations of asymmetric proximities, Research Report RR-91-10, University of Leiden, Department of Data Theory, 1991
20. M. Krawczak, G. Szkatuła, On asymmetric matching between sets. Inf. Sci. **312**, 89–103 (2015)
21. M. Krawczak, G. Szkatuła, Geometrical interpretation of impact of one set on another set, in *Proceedings 16th International Conference, ICAISC*, LNCS 10245, vol. 2017 (Springer, 2017), pp. 253–262
22. M. Krawczak, G. Szkatuła, Bidirectional comparison of multi-attribute qualitative objects. Inf. Sci. **436–437**, 367–387 (2018)
23. A. Tversky, *Preference, Belief, and Similarity. Selected Writings by Amos Tversky*. (Eldar Shafir, Massachusetts Institute of Technology, MIT Press, 2004)
24. J. Kacprzyk, M. Krawczak, G. Szkatuła, On bilateral matching between fuzzy set. Inf. Sci. **402**, 244–266 (2017)

Chapter 2
Issues of Asymmetry of Data Proximity

2.1 Preliminaries

In cognitive theory any process of comparison is fundamental and crucial for decision making, categorization, grouping, learning and so on. The study of the similarity of objects has been and still is of interest for philosophers as well as psychologists. Here, we consider a pair of objects A and B, for them a distance measure as well as the similarity can be calculated. However, when studying the similarity of objects often we are confused by the observation that the directional distance from the object A to object B can be different than the directional distance from the object B to object A. Such problems arise especially in cognitive and social psychology, in case of preferences of people (e.g., related to modeling of human similarity judgments), exchanges (e.g., import-export, brand switching), migration data, etc. It happens sometimes that when considering two objects one can notice that the object A is more associated with object B than the other way round. Asymmetry may have various meanings. Possible examples include also telephone calls between cities (e.g., the number of telephone calls from the city A to the city B can be different than the number of telephone calls from the city B to the city A).

The judgments depend on some distinguished features. Interesting argumentation about asymmetry of comparison of two objects was given by Rosch [1], concerning the case when one object is a prototype and the second a variant of the prototype. She considered objects like geometric figures or numbers.

The psychologist Amos Tversky in the paper *Features of similarity* [2] considered several data sets for which symmetry of similarity between the compared objects was not valid any more. Thus, he showed that judgments regarding similarity are in many cases not symmetrical. Then, he proposed to replace the geometrical approach to model similarity of objects by a set-theoretical approach, which became commonly used. His approach is called *the contrast model* and each object is characterized by distinguished features. This approach, the contrast theory, has become hugely influential over the past decades. For example, taking into account the "cost",

G. Szkatuła and M. Krawczak, *Bidirectional Comparison of Nominal Sets*,
Studies in Computational Intelligence 1140,
https://doi.org/10.1007/978-3-031-53096-8_2

in terms, say, of operations, of transformation of figures, figure "$\subseteq$" is more similar to figure "$\subset$", than figure "$\subset$" to figure "$\subseteq$". This way, judging the similarity, Tversky found that the less prominent stimulus was more similar to the prominent stimulus than the other way around (cf. Tversky [2]).

The objects can be viewed either as similar or as different, depending on the context and frame of reference (cf. Goodman [3]). Sometimes researchers perform some preprocessing of data to get symmetry. On the other hand, asymmetry may carry important information (e.g., Tversky [2, 4]; Tversky and Gati [5]; Tversky and Kahneman [6]). Thus, it seems that the assumption of symmetry should not be established in advance.

From yet another point of view, we can distinguish qualitative properties describing objects in subjective terms as well as quantitative properties describing objects in objective terms. The task of comparison of objects requires choosing proper methods of object feature representation as well as the computer data representation. In general, quantitative data represent numerical information about objects, and such information may be measured, for example $length$, $height$, $weight$, $time$, $cost$, etc. At the same time, qualitative data represent descriptive information about objects. Qualitative information is subjective and cannot be definitively measured. Thus, qualitative data can be observed but not measured, for example $beauty$, $smell$, $taste$, etc. In general, the qualitative data are described by sets of attributes and the attributes represented by nominal scales. Determination of similarities between "qualitative" objects by using common distance measures cannot be directly applicable for qualitative data. The problem of defining of proximity measures seems to be more difficult for nominal than for real-valued attributes.

There are several important issues where asymmetry of proximity of objects cannot be neglected, and even more, so the asymmetry determines the uniqueness of the considered phenomena. Therefore, here, let us recall some of them known from the literature, for instance the contrast model, the prospect theory, "salience" and "goodness" of the form, and "cost" of object transformation. The only assumption is that a measure of similarity or dissimilarity between two objects must be defined.

2.2 Feature Contrast Model

Amos Tversky proposed to consider similarity of objects by the theory of sets (i.e., sets of features of the considered objects), and it is exactly this approach that is widely exploited within this book. In 1977 Tversky [2] published a paper *Features of similarity* wherein the similarity of two entities increases with respect to their common features, and decreases with respect to their distinctive features. Common features can be understood as shared features, while distinctive features as non-shared features. Tversky proposed the following ternary models of similarity of two objects A and B:

$$similarity_1(A, B) = \frac{f(A \cap B)}{f(A \cap B) + \alpha \cdot f(A \setminus B) + \beta \cdot f(B \setminus A)}, \tag{2.1}$$

$$similarity_2(A, B) = \theta \cdot f(A \cap B) - \alpha \cdot f(A \setminus B) - \beta \cdot f(B \setminus A), \tag{2.2}$$

where A and B describe features of two considered objects, and f is a function of intersection $(A \cap B)$ or differences $(A \setminus B)$ or $(B \setminus A)$, and scalar parameters θ, α and β are introduced.

It can be said that the term $f(A \cap B)$ indicates common features of similarity, while the terms $\alpha \cdot f(A \setminus B)$ and $\beta \cdot f(B \setminus A)$ indicate distinctive features, namely the features which characterize one object and do not characterize the remaining object.

Such representations of similarity were named *the contrast model* since similarity is expressed as a relation (contrast) of the measures of the common features in reference to the distinctive features of two compared objects.

In the first *Tversky's contrast model* (called also *ratio contrast model*), Eq. 2.1, the similarity of two objects A and B is a function of common features for both objects (the numerator) divided by the sum of the common features increased by definite proportions of distinctive feature of the object A and distinctive features of the object B.

In the second *Tversky's contrast model*, Eq. 2.2, the similarity of two objects A and B is a function of common features for both objects reduced by definite proportions of distinctive features of the object A and the distinctive features of the object B. In these models, the parameters α, β and θ denote the weights of respective terms.

If we consider a non-directional problem, that is

"How similar are A and B?",

then the terms $\alpha \cdot f(A \setminus B)$ and $\beta \cdot f(B \setminus A)$ should be of the same importance, meaning $\alpha = \beta$. In this case, the similarity of two objects is symmetric.

Then, if we consider the directional problem

"How much A is like B"

then the term $\beta \cdot f(B \setminus A)$ should be weighed more heavily, and $\alpha < \beta$.

Considering the counterpart directional problem

"How much B is like A",

the term $\alpha \cdot f(A \setminus B)$ should be weighed more heavily (i.e., $\alpha > \beta$).

Additionally, it should be emphasized that the contrast models do not return the unique similarity scales. Thus, for $\alpha = \beta = 0$ the measure of similarity depends only on the shared features of compared two objects, for $\theta = 0$ the measure of similarity depends on the distinctive features of the two objects.

In general, if $f(A \setminus B) \neq f(B \setminus A)$ then the asymmetric relation between two sets takes place, that is

$$similarity_1(A, B) \neq similarity_1(B, A),$$

$$similarity_2(A, B) \neq similarity_2(B, A).$$

Applications of such measures of similarity of objects were illustrated by Tversky by the presentation of results of several experiments with evident asymmetry effects (e.g., comparisons of countries, geometrical figures and letters). Additionally, Tversky claims that many judgments of similarity have a naturally involved asymmetry, and then states that certain kinds of stimuli are more naturally used as focus than others (following Rosch [1]). Thus, in particular, following Tversky and Rosch, the more frequent item is a more likely focus, and so creates an asymmetric effect of similarity, which can be exemplified with respect to appearing of words in data bases (see Gawron [7]).

Within the problems of comparison of objects, there is another interesting question, namely the so called *self-similarity*. In mathematics, there is the term self-similarity, which is understood as a feature of a fractal whose parts are scaled and repeated. Here, in comparison of objects the term self-similarity has a different meaning. It is interesting to emphasize that applying different models of similarity we can obtain different results.

Thus, using the model of similarity (2.1) the self-similarity is equal one for any object. However, using the model (2.2), the self-similarity can be different for any object depending on parameter θ. Considering the distance models of similarity, the self-similarity of any object is the same [8].

2.3 Kahneman and Tversky: The Prospect Theory

The prospect theory was developed by Daniel Kahneman and Amos Tversky in 1979 [9] as a psychological model of human behavior in taking decisions under risk [10]. The theory has practical applications and describes behavioral economics and behavioral finance. Daniel Kahneman was awarded the 2002 Nobel Prize in Economics.

In outline, this theory describes human rationality in decisions involving risk. The theory states that people make decisions based on the potential value of losses and gains. The value function passes through the reference point (0,0), is S-shaped and asymmetrical (see Fig. 2.1).

The most evident characteristic of the prospect theory is that a magnitude of loss brings greater feeling of pain (convex value function) compared to the joy caused by an equivalent gain (concave value function). For example, see Fig. 2.1, the feeling of joy due to obtaining 100€ is lower than the pain caused by losing 100€.

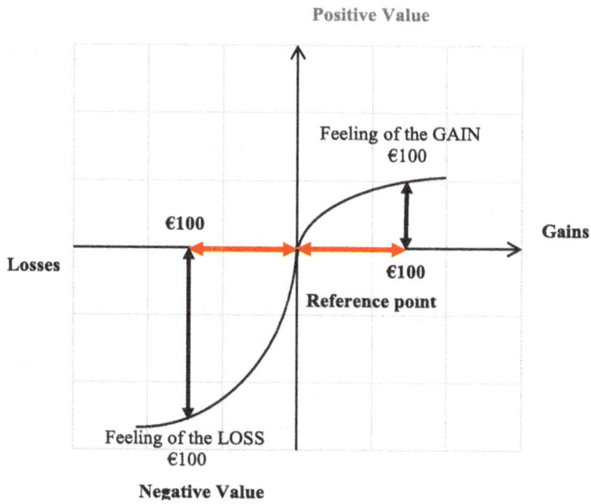

Fig. 2.1 A hypothetical value function [own study]

Perspective theory assumes that individuals make decisions based on expected loss or gain as perceived by their current wealth or welfare status. It is important to emphasize the statement that according to the prospect theory decisions are taken in reference to current status of wealth or welfare. Such comparisons to actual status is in some sense similar to the principles of the so called *Nash games* [11], wherein the Nash equilibrium is a solution of a non-cooperative game such that the players, when changing their strategies can only lose. John Nash was also awarded the 1994 Nobel in Economics.

In other words, the feature of the prospect theory is that more important are changes in wealth or welfare than the final outcomes. This theory is in contradiction to the *expected utility theory* (cf. [12]) where outcomes are maximized.

The essential remark concerning the prospect theory is the following: human beings are particularly averse to losing what they already have and less concerned with gains. Thus, the theory is another example of asymmetry, associated with the phenomenon of decision processes.

2.4 "Salience" and "Goodness" of Form

Another characteristic of Tversky's contrast model is the so called "salience" and "goodness" of form [2, 4, 13]. According to Tversky we consider objects as represented by sets of features, and measure object similarity via comparison of their common and distinctive features. Such assumptions generate many approaches to comparison of objects. Namely, when comparing two objects A and B we deal with the following fundamental questions:

"How similar is A to B?",

and

"How similar is B to A?".

The questions are directional and the similarity of the objects should not necessarily be a symmetric relation. For example, when comparing a person and his portrait (Tversky and Gati [5]), we say that

"The portrait resembles the person",

rather than

"The person resembles the portrait".

The perceived similarity is strictly associated with data representation. In general, the direction of asymmetry is determined by the relative "salience of the stimuli" [2]. Thus,

"The less salient stimulus is more similar to the more salient",

rather than

"The more salient stimulus is similar to the less salient".

If the object B is more salient than the object A, then A is more similar to B. In other words, a variant is more similar to the prototype than the prototype to a variant. A toy train is quite similar to a real train, because most features of the toy train are taken from the real train. On the other hand, a real train is not as similar to a toy train, because many of the features of a real train are not included in the toy train.

The psychological nature of human perception was discussed, in particular, by Tversky and Gati [5]. They hypothesized that both "goodness of form" and complexity contribute to the salience of geometric figures. Moreover, they expected that "good figure" be more salient than "bad figure". To investigate these hypotheses, they considered two sets of eight pairs of geometric figures. In the first set, one figure in each pair (denoted by p) has a "better" form than the other figure (denoted by q). In the second set, one figure in each pair (denoted by p) was "richer" or "more complex" than the other (denoted by q). Two figures from each of these sets are presented in Figs. 2.2 and 2.3.

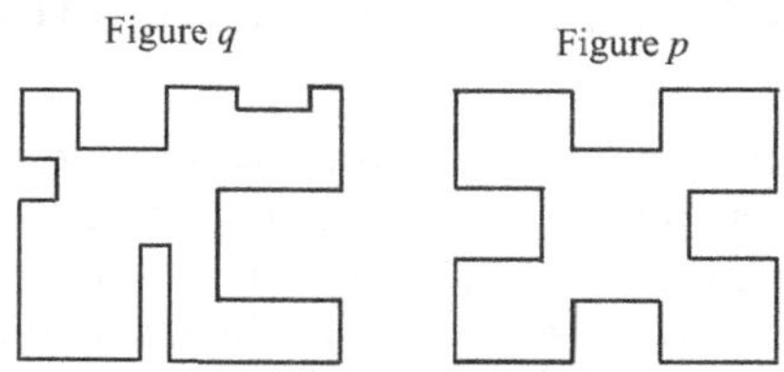

Fig. 2.2 A pair of figures from the first set, used to test the prediction of asymmetry [5]

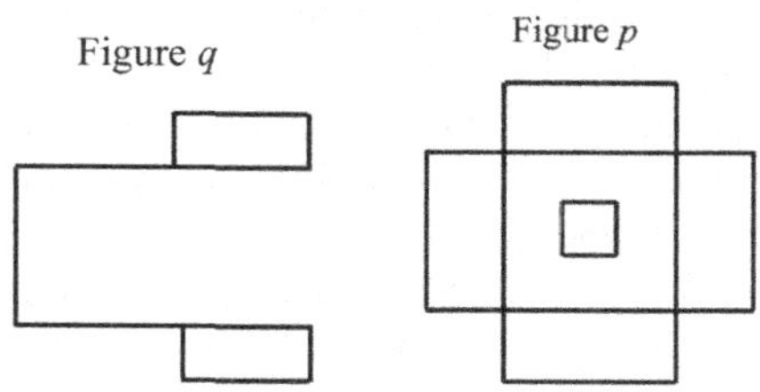

Fig. 2.3 A pair of figures from the second set, used to test the prediction of asymmetry [5]

In the experiment, a group of 69 respondents were involved and two elements of each pair were displayed to them side by side. The respondents were asked to choose one of the following two statements:

"The left figure is similar to the right figure",

or

"The right figure is similar to the left figure".

The order of the presented figures was randomized, so that figures appeared an equal number of times on the left as well as on the right side. In results, more than 2/3 of the respondents selected the form

"q is similar to p".

Within the second experiment, the same pairs of figures were used. One group of respondents was asked to estimate (on a 20-point scale) the degree to which the figure on the left was similar to the figure on the right, while the second group was asked to estimate the degree to which the figure on the right was similar to the figure on the left. In results, the hypothesis was confirmed that the average similarity of the figures q to the figures p, $S(q, p)$, was significantly higher than the average similarity of the figures p to the figures q, $S(p, q)$.

These experiments confirmed their hypothesis that similarity is asymmetrical, even if it does not clarify the concept of "goodness" of the form.

2.5 Transformational and Directional Distance

Hahn et al. [14] proposed the transformational approach to determining the similarity between objects. The approach is understood as addressing the complexity of transforming one object into another. It is obvious that similar objects require less steps of transformation, while dissimilar objects require more steps of transformation.

Thus, the number of simple steps required to transform one object into another can be considered as a directional distance, which can be asymmetric. Such distance is described by the minimal "cost" or "time" (the smallest number of elementary operations) of transformation by a computer program of the first object's representation to the second object's representation. Actually, this concept is known as the basis of the Levenshtein's distance [15].

We argue that the here proposed concept of measure of perturbation can be regarded as an extension of Levenshtein's distance. However, the concept of the set perturbation is evidently much more general because it is bidirectional and concerns nominal attributes.

However, there are some approaches which can be considered as a response to Tversky's critique of spatial models. Namely, Krumhansl [16] proposed *the distance-density model*, which is based on the spatial model and takes into account asymmetries. Next, Nosofsky [17], then proposed the use of differential stimulus bias to measure asymmetries within the spatial model. The bias is related to individual stimuli.

According to Tversky [2] as well as Garner and Haun [18], and Hahn, Chater and Richardson [14], the transformational approach assumes that similarity is determined by the complexity of "transforming" or "distorting" one object in order to arrive at the another one. It is obvious that similar objects require simple transformations while dissimilar objects require complex transformations. Hahn, Chater and Richardson [14] showed that transformational distance can be understood as the number of simple operations.

Such operations are, for example, additions and deletions of some features when comparing objects. It seems that deleting of some feature typically requires less complete specification compared to addition of them. Such transformations for the exemplary shapes (objects) *A* and *B* can be illustrated by Fig. 2.4. The task is to answer the questions:

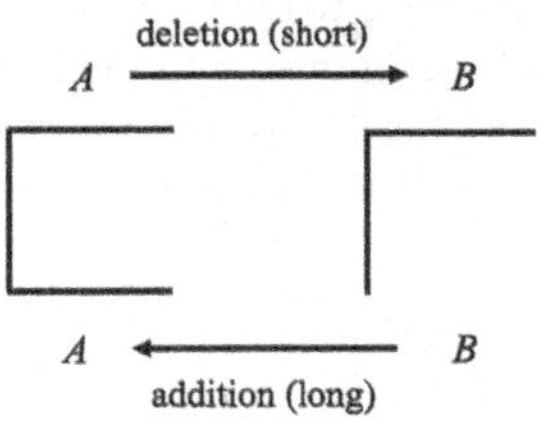

Fig. 2.4 Example of two shapes *A* and *B* [18]

"How similar is A to B?"

and

"How similar is B to A?",

where in the first case A is the target and B is the base, while in the second case B is the target and A is the base.

In order to generate the right figure from the left, the bottom line of the left figure should be deleted (the top arrow shows the direction of the transformation). In the opposite case, the process of adding bottom line is more complex because requires specification of "what" and "where" exactly to add (the down arrow shows the direction of the transformation). Having the difference between additions and deletions, Garner and Haun [18] noticed that "many-features" was more similar to "few-features" than vice versa. It is worth to mention that similar results were observed by Tversky [2].

One can also consider the "overall" transformation distance between two representations, which is characterized by the number of steps required to change one representation into another (e.g., Hodgetts et al. [19]). They distinguished three general transformations for comparing shapes:

- $create\ feature$ - taking the base object we apply this operation to create a new feature that is unique to the target object;
- $apply\ feature$ - this operation takes an object that is currently available (by being present in the base or by having been created via step 1) and applied to the base object;
- $swap$ - this swaps features between a pair of objects or swaps the object in its entirety.

In this way, considering the objects in Fig. 2.4, the transformation distance from left to right is 1, whereas it is 2 in the opposite direction. Considering complexity terms, the former transformation is simpler and therefore associated with greater perceived similarity.

Thus, the "overall" transformation distance between two representations is the number of steps required to turn one representation into another.

A different example is shown in Fig. 2.5, the transformation from the exemplary pair of two shapes - object A, to a different pair of two shapes - object B, and in the opposite direction.

Let us consider the first case. In order to calculate the transformation distance from the pair of shapes A to the pair of shapes B it is required to use only one transformation $apply$ for the existing $square$, that is

$$apply(square) = 1.$$

In the second case, the transformation distance from the pair of shapes B to the pair of shapes A requires using two transformations, $create$ a new triangle and $apply$ this new triangle, that is

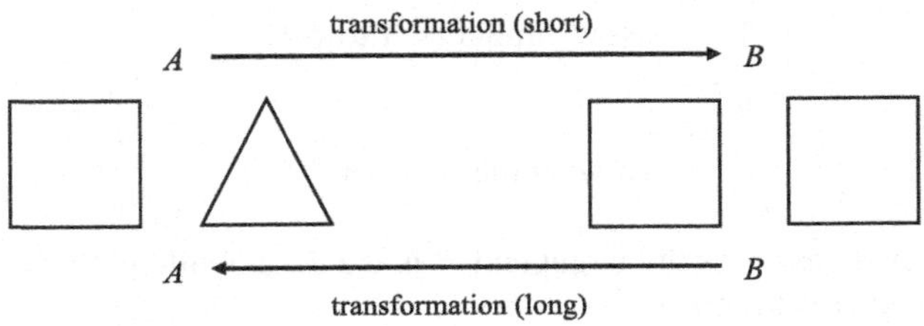

Fig. 2.5 Example of two pairs of two shapes A and B [19, 20]

$$create(triangle) + apply(triangle) = 2.$$

Thus, the transformation distance from the pair of two shapes A to the pair of two shapes B is "short" (requires one operation), whereas the transformation from the pair of two shapes B to the pair of two shapes A is "long" (requires two operations). The explanation is as follows: the application of the feature (square), that is currently available, is simpler than introducing the new feature (triangle).

Considering a non-directional similarity:

"How similar are A and B?",

the "overall" transformation distance for different shapes corresponds to the distance associated with the greater number of steps required to change one shape into another, which is the so called MAX-distance between the shapes (see Chater and Vitányi [21]).

Thus, transformations may occur to be more complicated in one direction than in another direction. Consequently, we face an asymmetry of object transformation, and therefore we can say that there is a different "cost" or "time" of transformation of one object into another than in the opposite case.

References

1. E. Rosch, Cognitive reference points. Cogn. Psychol. **7**(4), 532–547 (1975)
2. A. Tversky, Features of similarity. Psychol. Rev. **84**(4), 327–352 (1977)
3. N. Goodman, Seven strictures on similarity, in *Problems and Projects* (Bobs-Merril, New York, 1972), pp. 437–450
4. A. Tversky, in *Preference, Belief, and Similarity. Selected Writings by Amos Tversky*, ed. by E. Shafir (Massachusetts Institute of Technology, MIT Press, 2004)
5. A. Tversky, I. Gati, Studies of similarity, in *Cognition and Categorization*, ed. by E. Rosch, B. Lloyd, vol. 1 (Lawrence Elbaum Associates, 1978), pp. 79–98
6. A. Tversky, D. Kahneman, The framing of decisions and the psychology of choice. Science **211**, 453–458 (1981)
7. J. M. Gawron, Improving sparse word similarity models with asymmetric, in *Proceedings of the 52nd Annual Meeting of the Association for Computational Linguistics* (Baltimore, Maryland, USA, 23–25 June 2014), pp. 296–301

8. I. Gati, A. Tversky, Representations of qualitative and quantitative dimensions. J. Exp. Psychol. Hum. Percept. Perform. **8**(2), 325–340 (1982)
9. D. Kahneman, A. Tversky, Prospect theory: an analysis of decision under risk. Econometrica **47**(2), 263–292 (1979)
10. P. L. Bernstein, *Against the Gods: The Remarkable Story of Risk* (Wiley, 1996)
11. J. Nash, Non-cooperative games. Ann. Math. **54**(2), 286–295 (1951)
12. von J. Neumann, O. Morgenstern, *Theory of Games and Economic Behavior* (Princeton University Press, Princeton, NJ, 1944)
13. A. Tversky, D.H. Krantz, Similarity of schematic faces: a test of inter dimensional additivity. Percept. Psychophys. **5**, 124–128 (1969)
14. U. Hahn, N. Chater, L.B. Richardson, Similarity as transformation. Cognition **87**, 1–32 (2003)
15. V.I. Levenshtein, Binary codes capable of correcting deletions, insertions, and reversals. Sov. Phys. Doklady **10**, 707–710 (1966)
16. C.L. Krumhansl, Concerning the applicability of geometric models to similarity data: the interrelationship between similarity and spatial density. Psychol. Rev. **85**, 445–463 (1978)
17. R. Nosofsky, Stimulus bias, asymmetric similarity, and classification. Cognit. Psychol. **23**, 94–140 (1991)
18. W.R. Garner, F. Haun, Letter identification as a function of type of perceptual limitation and type of attribute. J. Exp. Psychol. Hum. Percept. Perform. **4**(2), 199–209 (1978)
19. C.J. Hodgetts, U. Hahn, N. Chater, Transformation and alignment in similarity. Cognition **113**, 62–79 (2009)
20. C.J. Hodgetts, U. Hahn, Similarity-based asymmetries in perceptual matching. Acta Psychol. **139**(2), 291–299 (2012)
21. N. Chater, P. Vitányi, The generalized universal law of generalization. J. Math. Psychol. **47**, 346–369 (2003)
22. J. Kacprzyk, M. Krawczak, G. Szkatuła, On bilateral matching between fuzzy set. Inf. Sci. **402**, 244–266 (2017)

Chapter 3
Matching Between Ordinary Sets

3.1 Preliminaries

In this chapter we propose a novel measure of proximity between two ordinary sets and our consideration is based on set-theoretic operations. Instead of considering distance between two ordinary sets, we introduce a definition of a measure of perturbation of one set on another set. The measure developed describes the changes of the first set after adding the second set or changes of the second set after adding the first set. The measure of set perturbation is normalized and returns a value from the interval [0, 1], where 1 is interpreted as the highest level of impact (perturbation), while 0 denotes the lowest level of impact. It must be emphasized that this measure is not symmetric, which means that the value of the measure of perturbation of the first set on the second set can be different then the value of the measure of perturbation of the second set on the first set. It is interesting that the sum of these measures can be regarded as an extended Jaccard's dissimilarity measure.

The reasoning is based on the fundamental properties of the classic set theory, and it is interesting that exploration of classic theories gives new interesting results. In addition, other interesting results were obtained, namely we could prove that several used measures of set similarity and dissimilarity can be described as proper functions of our impact measure. In this way we give some explanations of the nature of those measures as well as elegant clarifications. Further, it must be emphasized that the here introduced set impact measure allows for treating objects described by nominal-valued attributes in a direct way. Another approach to treating the nominal-valued attributes by binary encoding is described in the next Chapter.

Here, let us consider the finite set of objects $\{e_1, e_2, ..., e_N\}$. Then, we will consider each object e_n, $n \in \{1, 2, ..., N\}$, as described by a finite set of L nominal attributes.

Thus, a comparison of two objects can be based upon the values of such attributes. In general, a measure of similarity between objects is based on the compatibility of sets, each set being a description of an object. In this case simple matching leads

G. Szkatuła and M. Krawczak, *Bidirectional Comparison of Nominal Sets*,
Studies in Computational Intelligence 1140,
https://doi.org/10.1007/978-3-031-53096-8_3

to similarity expressed as a ratio of the number of matching attributes and the total number of attributes.

3.2 Examples of Proximity Measures of Ordinary Sets

Assume that we have a finite set V of nominal, distinct elements and a collection of subsets $\{A_1, A_2, ..., A_S\}$, where $A_1, A_2, ..., A_S \subseteq V$. Let us consider a pair of sets $A_i, A_j \subseteq V$, for $i, j \in \{1, 2, ..., S\}$.

For nominal elements, the comparison of one object with another one (i.e., one ordered set with another ordered set), can be considered in terms of whether the sets contain the same or different elements. Now, let us discus the following selected measures of set similarity or dissimilarity:

- *Jaccard's coefficient* (measure of similarity) and *Jaccard's distance* (measure of dissimilarity) are measures used to compare sets (e.g., Jaccard [1]; Shi [2]; Magurran [3]). Jaccard's coefficient for two sets, $S_{Jaccard}(A_i, A_j)$, is defined as the size of intersection of these two sets divided by the size of the union of these two sets:

$$S_{Jaccard}(A_i, A_j) = \frac{card(A_i \cap A_j)}{card(A_i \cup A_j)}. \tag{3.1}$$

The Jaccard coefficient is zero if two sets are disjoint, and is one if two sets are identical. On the other hand, Jaccard distance, $D_{Jaccard}(A_i, A_j)$, measures dissimilarity between two sets, and is complementary to the Jaccard coefficient as subtraction of the Jaccard coefficient from 1. Simple transformation shows that the Jaccard distance is equivalent to the difference of sizes of union and intersection of two sets divided by size of union of these two sets:

$$D_{Jaccard}(A_i, A_j) = 1 - S_{Jaccard}(A_i, A_j) =$$

$$= \frac{card(A_i \cup A_j) - card(A_i \cap A_j)}{card(A_i \cup A_j)}. \tag{3.2}$$

Here, we would like to note that the size of the common part of the sets $A_i^C \cap A_j^C$ is not included in the Jaccard coefficient, where A_i^C, A_j^C are the complements of sets A_i, A_j, in the set V, and therefore we can recall the following extended Jaccard's coefficient:

- *Jaccard's extended coefficient* takes into account the complements of sets A_i, A_j, in the set V. This coefficient is sometimes called *the simple matching coefficient* cf. [4], and can be written as follows:

$$\hat{S}_{Jaccard}(A_i, A_j) = \frac{card(A_i \cap A_j) + card(A_i^C \cap A_j^C)}{card(A_i \cup A_j) + card(A_i^C \cap A_j^C)} =$$

$$= \frac{card(A_i \cap A_j) + card(A_i^C \cap A_j^C)}{card(V)}. \tag{3.3}$$

where A_i^C, A_j^C are the complements of sets A_i, A_j, in the set V. It is important to notice that the Jaccard extended coefficient takes into account not only the elements belonging to both compared sets, but also the elements not belonging to these sets. In other words, the similarity of objects is affected not by only the common properties but also by the common gaps.

- *Dice's similarity coefficient*, $S_{Dice}(A_i, A_j)$ (also called the Sorensen coefficient) is characterized by normalization of intersection of two sets (e.g., Sorensen [5]; Dice [6]; Wolda [7]) and is shown below:

$$S_{Dice}(A_i, A_j) = \frac{2 \cdot card(A_i \cap A_j)}{card(A_i) + card(A_j)} =$$

$$= \frac{card(A_i \cap A_j)}{card(A_i \cap A_j) + \frac{1}{2} \cdot card(A_i \setminus A_j) + \frac{1}{2} \cdot card(A_j \setminus A_i)}. \tag{3.4}$$

The coefficient normalizes the intersection of two sets $A_i \cap A_j$ with the average of its constituents. The function ranges between zero and one, like in Jaccard's. Unlike Jaccard's distance, the corresponding difference function, based on Dice's similarity coefficient, namely $1 - S_{Dice}(A_i, A_j)$, is not a proper distance metric because it does not satisfy the triangle inequality property. It should be noticed that the following equalities are valid, namely

$$S_{Jaccard}(A_i, A_j) = \frac{S_{Dice}(A_i, A_j)}{2 - S_{Dice}(A_i, A_j)}$$

and

$$S_{Dice}(A_i, A_j) = \frac{2 \cdot S_{Jaccard}(A_i, A_j)}{1 + S_{Jaccard}(A_i, A_j)},$$

which means that these two are mutually monotonic.

Another interesting similarity coefficient is described below:

- *Overlap coefficient* (also called the Szymkiewicz-Simpson overlap coefficient)–this coefficient normalizes the intersection of two sets, $card(A_i \cap A_j)$, with the minimum cardinality of its arguments [8]:

$$S_{Ovl}(A_i, A_j) = \frac{card(A_i \cap A_j)}{min\{card(A_i), card(A_j)\}}. \tag{3.5}$$

We would like to pay special attention to Tversky's concept of proximity, which took its origin from psychology, namely:

- *Tversky's similarity* is the measure of degree to which two objects (viewed as sets of features) match each other (e.g., Tversky [9]). The matching between objects is expressed as a linear combination of the measures of common and distinctive features. The matching value is normalized to a value ranging from 0 to 1 and the formula used for this purpose (ratio contrast model (2.1)) is as follows:

$$S_{Tversky}(A_i, A_j; \alpha, \beta) = \frac{f(A_i \cap A_j)}{f(A_i \cap A_j) + \alpha \cdot f(A_i \setminus A_j) + \beta \cdot f(A_j \setminus A_i)}. \tag{3.6}$$

for some parameters $\alpha, \beta > 0$.

The values of α and β determine he relative importance of the distinctive features in the similarity assessment. The similarity is based on a function $f(.)$ called the measure of sets. A finite set can be measured by the number of elements (i.e., the cardinality of a set), but may be measured by any function that satisfies the property of additivity (i.e., is a function satisfying the condition $f(A_i \cup A_j) = f(A_i) + f(A_j)$ for the disjoint sets A_i and A_j).

It seems that an interesting point of the object similarity consideration is to treat the first object e_1 as the prototype and the second object e_2 as the variant, and then parameter α corresponds to the weight of the prototype while parameter β corresponds to the weight of the variant. The discussion, related to participation of prototype and variant in Eq. 3.6 can be very interesting, but we would like to emphasize, for example, the case when the weighting of prototype features is equal to 100% (i.e., $\alpha = 1$) and of the variant features to 0% (i.e., $\beta = 0$)—mining that only the prototype features are considered as important.

Taking into consideration the common part of the sets $A_i^C \cap A_j^C$, where A_i^C and A_j^C are the complements of sets A_i and A_j in the set V, we can used the following Tversky's extended similarity measure:

- *Tversky's extended similarity measure* between the sets A_i and A_j can be written in the following form:

$$\hat{S}_{Tversky}(A_i, A_j; \alpha, \beta) =$$

$$= \frac{card(A_i \cap A_j) + card(A_i^C \cap A_j^C)}{card(A_i \cap A_j) + \alpha \cdot card(A_i \setminus A_j) + \beta \cdot card(A_j \setminus A_i) + card(A_i^C \cap A_j^C)}. \tag{3.7}$$

where A_i^C, A_j^C are the complements of the sets A_i and A_j in the set V.

It is worth to notice that Tversky's extended similarity measure for parameters $\alpha = \beta = 1$ becomes just the Jaccard extended coefficient.

For an exemplary set V and two disjoint subsets $A_i, A_j \subseteq V$ (i.e., the condition $A_i \cap A_j = \emptyset$ is fulfilled), and according to Eqs. 3.1, 3.4, 3.5 and 3.6, we obtain the following values of the measures of similarity:

$$S_{Jaccard}(A_i, A_j) = 0, \; S_{Dice}(A_i, A_j) = 0, \; S_{Ovl}(A_i, A_j) = 0, \text{ and } S_{Tversky}(A_i, A_j) = 0.$$

3.3 Basics of Perturbation of Ordinary Sets

Let us consider a finite set V of nominal, distinct values, $V = \{v_1, v_2, \ldots, v_L\}$, and a collection of subsets $\{A_1, A_2, ..., A_S\}$, where $A_1, A_2, ..., A_S \subseteq V$.

If consecutive values are labeled by the letters of the alphabet, we can describe an exemplary set V as for example $\{a, b, c, d, e, f, g\}$; or when they are labeled by words, we can describe an exemplary set V as, for instance, $\{salty, sweet, sour, bitter, tasteless\}$.

3.3.1 *The Concept of Perturbation of Ordinary Sets*

Let us consider a pair of sets $A_i, A_j \subseteq V$, for $i, j \in \{1, 2, ..., S\}$. Let us attach the first set A_i to the second set A_j. Such attachment can be perceived so that the first set impacts on the second set, or, in other words, the second set is perturbed by the first set, or, in other words the set A_i perturbs the set A_j to some degree.

In this way, we introduce the concept of *the perturbation of ordinary set* A_j *by ordinary set* A_i, denoted by $(A_i \mapsto A_j)$, which is interpreted as the relative complement of A_i in A_j (also called the set-theoretic difference of A_i and A_j), denoted by $A_i \setminus A_j$.

For example, the set $A_1 = \{e\}$ perturbs the set $A_2 = \{a, b, c, d, e\}$ with zero degree, because the following condition is satisfied

$$(A_1 \mapsto A_2) = A_1 \setminus A_2 = \emptyset.$$

On the other hand, the set A_2 perturbs the set A_1 with the degree greater than zero, because

$$(A_2 \mapsto A_1) = A_2 \setminus A_1 = \{a, b, c, d\}.$$

3.3.2 *Measures of Perturbation of Ordinary Sets*

After normalization, the measure of such perturbation of sets lies within the range from 0 to 1, where 1 is the highest value of perturbation, while 0 is the lowest value of perturbation. Notice that the measure is zero only if the first set constitutes a subset

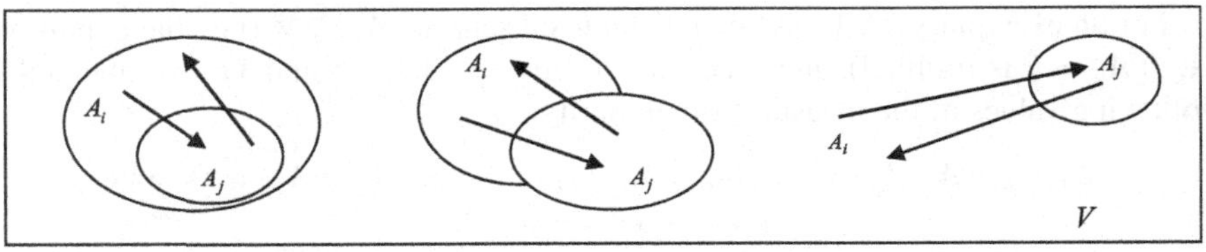

Fig. 3.1 Perturbations of one ordinary set by another

of the second set. For disjoint sets the measure of perturbation is greater than zero. In Fig. 3.1 three main cases are shown of reciprocal relationships between two sets, that is $A_j \subseteq A_i$, $A_i \cap A_j \neq \emptyset$ and $A_i \cap A_j = \emptyset$, with the directions of perturbation (impact) being indicated by arrows.

The measure of impact (perturbation) can be normalized in different ways. Here, we recall two types of the measures of perturbation of one set by another, denoted by perturbation of type 1 and perturbation of type 2 (cf. Krawczak and Szkatuła [10, 11]). However, we will consider perturbation of types 1 and type 2 interchangeably.

In this book we propose the measure of perturbation of one set by another, as in Definition 3.1, and this kind of set perturbation will be called perturbation of type 1.

Definition 3.1 (*Perturbation measure of type 1*) The measure of the perturbation of type 1 of the set A_j by the set A_i is denoted by $Per^1_{OS}(A_i \mapsto A_j)$, and is defined in the following manner:

$$Per^1_{OS}(A_i \mapsto A_j) =$$

$$= \frac{card(A_i \setminus A_j)}{card(A_i \cap A_j) + card(A_i \setminus A_j) + card(A_j \setminus A_i) + card(A_i^C \cap A_j^C)} =$$

$$= \frac{card(A_i \setminus A_j)}{card(V)}. \tag{3.8}$$

where A_i^C, A_j^C are the complements of the sets $A_i, A_j \subseteq V$.

The measure of perturbation of type 2 also may be introduced, and in this case the perturbation is normalized by the size of the union of these two sets.

Definition 3.2 (*Perturbation measure of type 2*) The measure of the perturbation of type 2 of the set A_j by the set A_i is denoted by $Per^2_{OS}(A_i \mapsto A_j)$, and is defined in the following manner:

$$Per^2_{OS}(A_i \mapsto A_j) = \frac{card(A_i \setminus A_j)}{card(A_i \cap A_j) + card(A_i \setminus A_j) + card(A_j \setminus A_i)} =$$

$$= \frac{card(A_i \setminus A_j)}{card(A_i \cup A_j)}. \tag{3.9}$$

The idea of perturbation can now be illustrated by the following example.

Example 3.1 Assume that we have an exemplary set $V = \{a, b, c, d, e\}$ and two subsets $A_1, A_2 \subseteq V$, where

$$A_1 = \{a, b, c\} \text{ and } A_2 = \{b\}.$$

Using the Definition 3.1 we obtain the following values of the measure of perturbation of type 1 of one set by another, namely:

$$Per^1_{OS}(A_1 \mapsto A_2) = 2/5 \text{ and } Per^1_{OS}(A_2 \mapsto A_1) = 0. \qquad \square$$

3.3.3 Properties of Measures of Ordinary Set Perturbation

First, we can consider the following corollary about the minimum value of the measure of perturbation of set A_j by set A_i, this value being equal zero (e.g., Krawczak and Szkatuła [11]).

Corollary 3.1 *The measure of perturbation of type 1 of the set A_j by the set A_i, satisfies the following property*

$$Per^1_{OS}(A_i \mapsto A_j) = 0 \text{ if and only if } A_i \subseteq A_j. \tag{3.10}$$

Proof (1) First, we prove the left hand side implication:

$$Per^1_{OS}(A_i \mapsto A_j) = 0 \Rightarrow A_i \subseteq A_j.$$

Let us assume that $Per^1_{OS}(A_i \mapsto A_j) = 0$. By Definition 3.1, function Per^1_{OS} $(A_i \mapsto A_j)$ is non negative, and reaches a minimum if the condition $card(A_i \setminus A_j) = 0$ is satisfied. If $card(A_i \setminus A_j) = 0$, then condition $A_i \subseteq A_j$ is fulfilled.

(2) Consider now the implication:

$$A_i \subseteq A_j \Rightarrow Per^1_{OS}(A_i \mapsto A_j) = 0.$$

Let us assume that $A_i \subseteq A_j$, thus $A_i \setminus A_j = \emptyset$ and $card(A_i \setminus A_j) = 0$. Therefrom, we obtain that $Per^1_{OS}(A_i \mapsto A_j) = 0$, by Definition 3.1. The equality $Per^1_{OS}(A_i \mapsto A_j) = 0$ is always verified if $A_i \subseteq A_j$. $\square$

It should be noticed that the measure of perturbation of type 1 of set A_j by set A_i is not always symmetrical (i.e., $Per^1_{OS}(A_i \mapsto A_j) \neq Per^1_{OS}(A_j \mapsto A_i)$). The measure is symmetrical whenever the condition

$$card(A_i \mapsto A_j) = card(A_j \mapsto A_i)$$

is satisfied. We can say that the asymmetry is determined by the relative cardinality of the sets, that is $Per^1_{OS}(A_i \mapsto A_j) \geq Per^1_{OS}(A_j \mapsto A_i)$ whenever the inequality $card(A_i \mapsto A_j) \geq card(A_j \mapsto A_i)$ is fulfilled.

We get similar dependencies for the measure of perturbation of type 2 [11].

Corollary 3.2 *The measure of perturbation of type 2 of the set* A_j *by the set* A_i*, satisfies the following property*

$$Per^2_{OS}(A_i \mapsto A_j) = 0 \ if \ and \ only \ if \ \ A_i \subseteq A_j. \tag{3.11}$$

Proof By Definition 3.2. □

Yet other properties relating to the range of the measures can be proved, namely the measure of the set perturbation is always positive and less than 1, where 1 is interpreted as the highest level of perturbation, while 0 is the lowest level of perturbation, as shown in Corollaries 3.3 and 3.4 [11].

Corollary 3.3 *The measure of perturbation of type 1 of the set* A_j *by the set* A_i*, satisfies the following inequality*

$$0 \leq Per^1_{OS}(A_i \mapsto A_j) \leq 1. \tag{3.12}$$

Proof (1) Let us prove the first inequality $Per^1_{OS}(A_i \mapsto A_j) \geq 0$. It should be noticed that the inequality $card(A_i \mapsto A_j) \geq 0$ is satisfied, and by Definition 3.1 we obtain $Per^1_{OS}(A_i \mapsto A_j) \geq 0$.

(2) Now, we will consider the second inequality, $Per^1_{OS}(A_i \mapsto A_j) \leq 1$. Considering two sets A_i and A_j, where $A_i, A_j \subseteq V$, it should be noticed that the inequality $card(A_i \setminus A_j) \leq card(V)$ is satisfied, and then we can obtain the following inequality

$$Per^1_{OS}(A_i \mapsto A_j) = \frac{card(A_i \setminus A_j)}{card(V)} \leq 1.$$

□

We get similar dependencies for the measure of perturbation of type 2.

Corollary 3.4 *The measure of perturbation of type 2 of the set* A_j *by the set* A_i*, satisfies the following inequality*

$$0 \leq Per^2_{OS}(A_i \mapsto A_j) \leq 1. \tag{3.13}$$

Proof (1) Let us prove the first inequality $Per^2_{OS}(A_i \mapsto A_j) \geq 0$. It should be noticed that the inequality $card(A_i \mapsto A_j) \geq 0$ is satisfied, and by Definition 3.2 we obtain $Per^2_{OS}(A_i \mapsto A_j) \geq 0$.

(2) Now, we will consider the second inequality, $Per^2_{OS}(A_i \mapsto A_j) \leq 1$. Considering two sets A_i and A_j, where $A_i, A_j \subseteq V$, it should be noticed that the inequality

$card(A_i \setminus A_j) \le card(A_i \cup A_j))$ is satisfied, and then we can obtain the following inequality

$$Per^2_{OS}(A_i \mapsto A_j) = \frac{card(A_i \setminus A_j)}{card(A_i \cup A_j)} \le 1.$$

□

Here, we will prove a very interesting property of the sum of measures of perturbation of arbitrary two sets, namely perturbation of set A_i by set A_j and perturbation of set A_j by set A_i in the in the following corollary (cf. [11]).

Corollary 3.5 *The sum of the measures of perturbation of type 1 of the arbitrary sets A_j and A_i satisfies the following inequality*

$$0 \le Per^1_{OS}(A_i \mapsto A_j) + Per^1_{OS}(A_j \mapsto A_i) \le 1 \tag{3.14}$$

Proof (1) Due to Corollary 3.1, the sum of the following perturbations

$$Per^1_{OS}(A_i \mapsto A_j) + Per^1_{OS}(A_j \mapsto A_i)$$

is non negative.

(2) It can be noticed that the inequalities

$$card(A_i \setminus A_j) + card(A_j \setminus A_i) \le card(A_i \cup A_j), \;\; and$$

$$card(A_i \cup A_j) \le card(V)$$

are satisfied. So, the following inequality

$$\frac{card(A_i \setminus A_j)}{card(V)} + \frac{card(A_j \setminus A_i)}{card(V)} \le \frac{card(A_i \cup A_j)}{card(V)} \le \frac{card(V)}{card(V)} = 1$$

are fulfilled. □

Additionally, we can prove an interesting property of the here introduced measure of perturbation of type 1 of one set by another set, and the Jaccard extended coefficient for these sets, presented as Corollary 3.6.

Corollary 3.6 *The sum of the measures of perturbation of type 1 of arbitrary set A_j and the set A_i satisfies the following equality*

$$Per^1_{OS}(A_i \mapsto A_j) + Per^1_{OS}(A_j \mapsto A_i) = 1 - \hat{S}_{Jaccard}(A_i, A_j) \tag{3.15}$$

where $\hat{S}_{Jaccard}(A_i, A_j)$ is Jaccard's extended coefficient for the sets A_i and A_j, according to (3.7).

Proof By Definition 3.1 the left hand side of Eq. (3.15) can be rewritten as follows

$$Per^1_{OS}(A_i \mapsto A_j) + Per^1_{OS}(A_j \mapsto A_i) = \frac{card(A_i \setminus A_j) + card(A_j \setminus A_i)}{card(V)}.$$

By adding and subtracting the following expression in the numerator

$$card(A_i \cap A_j) + card(A_i^C \cap A_j^C)$$

we obtain

$$1 - \frac{card(A_i \cap A_j) + card(A_i^C \cap A_j^C)}{card(V)} = 1 - \hat{S}_{Jaccard}(A_i, A_j).$$

□

Formula (3.15) has a very interesting novel interpretation. Namely, the term $\hat{S}_{Jaccard}(A_i, A_j)$ is defined by (3.3) as Jaccard's extended coefficient of set similarity through the use of the complements of the sets A_i and A_j. Then, the term

$$1 - \hat{S}_{Jaccard}(A_i, A_j)$$

can be considered as Jaccard's extended measure of dissimilarity of sets. In this way the formula (3.15) describes an equivalent definition of the Jaccard extended measure of dissimilarity of sets, that is

$$1 - \hat{S}_{Jaccard}(A_i, A_j) = Per^1_{OS}(A_i \mapsto A_j) + Per^1_{OS}(A_j \mapsto A_i).$$

Therefore, it can be said that the sum of two coupled measures of perturbation of sets, provides an equivalent interpretation of dissimilarity of these sets.

Example 3.2 Let us consider the exemplary set $V = \{b, c, d, e, f, m, n\}$ and two subsets $A_1, A_2 \subseteq V$, where

$$A_1 = \{b, c, d\} \text{ and } A_2 = \{m, b, c, e, f\}.$$

The perturbation measures between these two sets and the Jaccard extended coefficient are calculated the following:

$$Per^1_{OS}(A_1 \mapsto A_2) = \frac{1}{7}, \quad Per^1_{OS}(A_2 \mapsto A_1) = \frac{3}{7}, \quad \hat{S}_{Jaccard}(A_1 \mapsto A_2) = \frac{3}{7},$$

hence, formula (3.15) is obviously satisfied. □

Then, similar dependencies can be proved for the measure of perturbation of type 2.

Corollary 3.7 *The sum of the measures of perturbation of type 2 for arbitrary sets A_j and A_i satisfies the following inequality*

$$0 \leq Per^2_{OS}(A_i \mapsto A_j) + Per^2_{OS}(A_j \mapsto A_i) \leq 1 \tag{3.16}$$

Proof (1) Due to Definition 3.2, the sum of the following perturbations

$$Per^2_{OS}(A_i \mapsto A_j) + Per^2_{OS}(A_j \mapsto A_i)$$

is non negative.

(2) It can be noticed that the inequality

$$card(A_i \setminus A_j) + card(A_j \setminus A_i) \leq card(A_i \cup A_j)$$

is satisfied. The right hand side of the above inequality can be written as

$$\frac{card(A_i \setminus A_j)}{card(A_i \cup A_j)} + \frac{card(A_j \setminus A_i)}{card(A_i \cup A_j)} \leq \frac{card(A_i \cup A_j)}{card(A_i \cup A_j)} = 1.$$

□

3.4 Representation of Multi-attribute Objects by Ordinary Sets

Let us consider a collection of the multi-attribute qualitative objects $U = \{e_n\}$, indexed by n, for $n = 1, 2, ..., N$. The objects are described by K attributes

$$A = \{a_1, a_2, ..., a_K\},$$

indexed by j, for $j = 1, 2, ..., K$. The set

$$V_{a_j} = \{v_{1,j}, v_{2,j}, ..., v_{L_j,j}\}$$

is the domain of the attribute $a_j \in A$, $j = 1, 2, ..., K$, where L_j denotes the number of nominal values of the attribute a_j, $L_j \geq 1$.

3.4.1 Description of Multi-attribute Objects by Ordinary Sets

Assuming that the objects are represented by their descriptions, *the description of an object* $e_n \in U$ is denoted by G_{e_n}, and can be represented by an ordered collection of ordinary sets, see the following definition (cf. Krawczak and Szkatuła [11]).

Definition 3.3 (*Description of object*) Every object e_n, $e_n \in U$, can be represented by an ordered collection of K singletons, (i.e., K ordered values) $v_{j,t(j,e_n)}$, $j \in \{1, 2, ..., K\}$, drawn from the ordinary sets of nominal elements $V_{a_j} = \{v_{1,j}, v_{2,j}, ..., v_{L_j,j}\}$ of the attributes a_j, described as follows

$$G_{e_n} =< A_{1,t(1,e_n)}, A_{2,t(2,e_n)}, ..., A_{K,t(K,e_n)} >=$$

$$=< \{v_{1,t(1,e_n)}\}, \{v_{2,t(2,e_n)}\}, ..., \{v_{K,t(K,e_n)}\} >, \tag{3.17}$$

where $v_{j,t(j,e_n)} \in V_{a_j}$, $j \in \{1, 2, ..., K\}$.

The index $t(j, e_n)$, $j \in \{1, 2, ..., K\}$ and $n \in \{1, 2, ..., N\}$, denotes that the attribute a_j takes the value $v_{j,t(j,e_n)}$ for the object e_n.

The arithmetic addition (marked by $\oplus$) can be applied to all sets in the descriptions of the objects, for corresponding attributes. In this way can introduce the definition of the join between the descriptions of the objects.

Definition 3.4 (*Join between descriptions of objects*) The join between the description of an object e_1 and the description of an object e_2 is defined as follows

$$G_{e_1} \oplus G_{e_2} =< (A_{1,t(1,e_1)} \oplus A_{1,t(1,e_2)}), ..., (A_{K,t(K,e_1)} \oplus A_{K,t(K,e_2)}) > . \tag{3.18}$$

The definition says that the description of two joined objects is again a collection of K sets. Each j-th set, $j = \{1, 2, ..., K\}$, is constructed as the join of two sets $A_{j,t(j,e_1)} \oplus A_{j,t(j,e_2)}$, describing the attribute a_j for the objects e_1 and e_2, respectively.

3.4.2 Measure of Perturbation of Multi-attribute Objects

Now, let us consider two objects, $e_1, e_2 \in U$, and each object can be represented by an ordered collection of sets of values of the attributes, describing objects (i.e., G_{e_1} and G_{e_2}, respectively), as follows

$$G_{e_1} =< A_{1,t(1,e_1)}, A_{2,t(2,e_1)}, ..., A_{K,t(K,e_1)} >,$$

$$G_{e_2} =< A_{1,t(1,e_2)}, A_{2,t(2,e_2)}, ..., A_{K,t(K,e_2)} > .$$

The arithmetic subtraction (marked by $\ominus$) can be applied to all sets in the descriptions of the objects, for corresponding attributes. In this way we can introduce the definition of the perturbation between the descriptions of the objects. The concept of object's perturbation is defined as follows (cf. Krawczak and Szkatuła [11]).

Definition 3.5 (*Perturbation of objects*) The perturbation of the object e_2 by the object e_1, denoted by $(G_{e_1} \mapsto G_{e_2})$, can be represented by an ordered collection of sets $A_{j,t(j,e_1)} \ominus A_{j,t(j,e_2)}$, $j = 1, 2, ..., K$, drawn from the ordinary sets of nominal values V_{a_j} of the attributes a_j, respectively

$$(G_{e_1} \mapsto G_{e_2}) =< (A_{1,t(1,e_1)} \mapsto A_{1,t(1,e_2)}), ..., (A_{K,t(K,e_1)} \mapsto A_{K,t(K,e_2)}) >=$$

$$=< A_{1,t(1,e_1)} \ominus A_{1,t(1,e_2)}, A_{2,t(2,e_1)} \ominus A_{2,t(2,e_2)}, ..., A_{K,t(K,e_1)} \ominus A_{K,t(K,e_2)} > \quad (3.19)$$

Thus, the perturbation of the object e_2 by the object e_1 is represented by the collection of the sets constructed as differences of the sets, for each attribute a_j, $j = 1, 2, ..., K$.

The counterpart case is defined in a similar way, that is

$$(G_{e_2} \mapsto G_{e_1}) =< (A_{1,t(1,e_2)} \mapsto A_{1,t(1,e_1)}), ..., (A_{K,t(K,e_2)} \mapsto A_{K,t(K,e_1)}) >=$$

$$=< (A_{1,t(1,e_2)} \ominus A_{1,t(1,e_1)}), (A_{2,t(2,e_2)} \ominus A_{2,t(2,e_1)}), ..., (A_{K,t(K,e_2)} \ominus A_{K,t(K,e_1)}) > . \quad (3.20)$$

In turn, *the measure of the perturbation* of the one object by another object is a number ranging between 0 and 1 and obtained via some aggregation operation. The aggregation is done on a set of measures of the perturbations, associated with each attribute a_j, $j = 1, 2, ..., K$, see Definition 3.6 [11].

Definition 3.6 (*Measure of perturbation of objects*) The measure of perturbation of the object e_2 by the object e_1, denoted generally by $Per_{OS}(G_{e_1} \mapsto G_{e_2})$, is defined in the following manner:

$$Per_{OS}(G_{e_1} \mapsto G_{e_2}) =$$

$$= Agg(Per_{OS}(A_{1,t(1,e_1)} \mapsto A_{1,t(1,e_2)}), ..., Per_{OS}(A_{K,t(K,e_1)} \mapsto A_{K,t(K,e_2)})) \quad (3.21)$$

where $Per_{OS}(A_{j,t(j,e_1)} \mapsto A_{j,t(j,e_2)})$, for $j = 1, 2, ..., K$, is defined according to Eq. 3.8, and Agg is an aggregation operation.

The aggregation operation used in (3.21) is defined as a mapping

$$Agg : [0, 1]^K \to [0, 1],$$

that assigns a real number p to any K-tuple $(p_1, p_2, ..., p_K)$ of real numbers, and satisfies the following conditions:

- *idempotence*: $Agg(p, p, ..., p) = p$;
- *monotonicity*:
  if $p_i \geq q_i$ for $i = 1, 2, ..., K$, then $Agg(p_1, p_2, ..., p_K) \geq Agg(q_1, q_2, ..., q_K)$;
- *boundary conditions*: $Agg(0, 0, ..., 0) = 0$ and $Agg(1, 1, ..., 1) = 1$;
- *commutativity*: $Agg(p_1, p_2, ..., p_K) = Agg(p_{i_1}, p_{i_2}, ..., p_{i_K})$, for every permutation $(i_1, i_2, ..., i_K)$ of (1, 2, ...,K).

In general, it is assumed that the result of the aggregation is not higher than the highest element aggregated (the maximum) and is not lower than the lowest one (the minimum) [12], that is, the following inequalities

$$\min_{j \in \{1,2,...,K\}} \{p_j\} \leq Agg(p_1, p_2, ..., p_K) \leq \max_{j \in \{1,2,...,K\}} \{p_j\}$$

are satisfied.

The aggregation operation Agg can be realized by various functions, for example

- *minimum*: $Agg(p_1, p_2, ..., p_K) := min\{p_1, p_2, ..., p_K\}$;
- *maximum* : $Agg(p_1, p_2, ..., p_K) := max\{p_1, p_2, ..., p_K\}$;
- *arithmetic average*: $Agg(p_1, p_2, ..., p_K) := \frac{1}{K} \sum_{j=1}^{K} (p_j)$;
- *weighted average*: $Agg(p_1, p_2, ..., p_K) := \frac{1}{K} \sum_{j=1}^{K} (w_j \cdot p_j)$;
- *generalized arithmetic mean*: $Agg(p_1, p_2, ..., p_K) := (\frac{1}{K} \sum_{j=1}^{K} (p_j^{\alpha}))^{\frac{1}{\alpha}}$.

Let us assume that the weight parameter $w_j \geq 0$ determines *the importance* of the jth element, for $j = 1, 2, ..., K$. In the further considerations it is assumed that the aggregation operation Agg can be realized by a function of weighted average of its arguments.

3.4.3 Measure of Perturbation of Groups of Multi-attribute Objects

Let us assume that every non empty subset of a finite set of objects U is called a *group*. Each group $g \subseteq U$ can be represented by *a description of a group* as an ordered collection of K sets $A_{j,t(j,g)}$, for $j = 1, 2, ..., K$, drawn from the ordinary sets of values

$$V_{a_j} = \{v_{1,j}, v_{2,j}, ..., v_{L_j,j}, \}$$

of the attributes describing objects a_j, in other words

$$G_g =< A_{1,t(1,g)}, A_{2,t(2,g)}, ..., A_{K,t(K,g)} >, \tag{3.22}$$

where $A_{j,t(j,g)} \subseteq V_{a_j}$ and $card(A_{j,t(j,g)}) \leq L_j$, for $j \in \{1, 2, ..., K\}$.

Now, let us consider two non-empty groups $g_1 \in U, g_2 \in U$, where each group can be represented by an ordered collection of sets of values of the attributes describing objects (i.e., G_{g_1} and G_{g_2}), as follows

$$G_{g_1} =< A_{1,t(1,g_1)}, A_{2,t(2,g_1)}, ..., A_{K,t(K,g_1)} >,$$

$$G_{g_2} =< A_{1,t(1,g_2)}, A_{2,t(2,g_2)}, ..., A_{K,t(K,g_2)} > .$$

The arithmetic addition of the sets is a new set, and can be applied to all sets in the descriptions G_{g_1} and G_{g_2}. In this way we can introduce a definition of the join between the descriptions of groups (cf. Krawczak and Szkatuła [11, 13, 14]).

Definition 3.7 (*Join between descriptions of groups*) The join between the description of group G_{e_1} and the description of group G_{e_2} is defined as follows

$$G_{g_1} \oplus G_{g_2} =< (A_{1,t(1,g_1)} \oplus A_{1,t(1,g_2)}), ..., (A_{K,t(K,g_1)} \oplus A_{K,t(K,g_2)}) > . \tag{3.23}$$

The definition says that the description of two joined groups is again a collection of K sets. Each such jth set, $j = 1, 2, ..., K$, is constructed as the join of two sets $A_{j,t(j,g_1)} \oplus A_{j,t(j,g_2)}$ describing the attribute a_j for the groups g_1 and g_2, respectively.

According to the measure of perturbation of sets, Eq. 3.8, we will introduce the definition of perturbation measure for two groups. The definition concerns a separate attribute a_j, $j \in \{1, 2, ..., K\}$.

Definition 3.8 (*Measure of perturbation for groups of objects*) The measure of the perturbation of the group G_{g_2} by the group G_{g_1} (denoted generally by $Per_{OS}(G_{g_1} \mapsto G_{g_2})$, is defined in the following manner:

$$Per_{OS}(G_{g_1} \mapsto G_{g_2}) =$$

$$= Agg\big(Per_{OS}(A_{1,t(1,g_1)} \mapsto A_{1,t(1,g_2)}), ..., Per_{OS}(A_{K,t(K,g_1)} \mapsto A_{K,t(K,g_2)})\big) \tag{3.24}$$

where Agg is the aggregation operation.

The aggregation operation used in (3.24) is defined as a mapping $Agg : [0, 1]^K \to [0, 1]$, which assigns a real number p to any K-tuple $(p_1, p_2, ..., p_K)$ of real numbers, see Sect. 3.4.2.

3.5 Illustrative Examples

3.5.1 Description of Group of Objects

The meaning of (3.23) can be illustrated by a simple example.

Let us consider seven objects

$$U = \{e_1, e_2, e_3, e_4, e_5, e_6, e_7\},$$

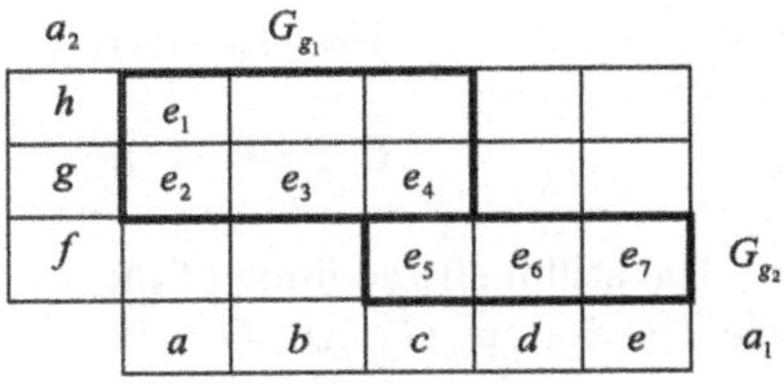

Fig. 3.2 Exemplary groups $g_1 = \{e_1, e_2, e_3, e_4\}$ and $g_2 = \{e_5, e_6, e_7\}$, represented by G_{g_1} and G_{g_2}, respectively

described by two attributes $\{a_1, a_2\}$ and the attributes have the following domains

$$V_{a_1} = \{a, b, c, d, e\} \text{ and } V_{a_2} = \{f, g, h\}.$$

Let us assume that the objects are represented according to (3.22) as follows:

$$e_1 =< \{a\}, \{h\} >, \ e_2 =< \{a\}, \{g\} >, \ e_3 =< \{b\}, \{g\} >, \ e_4 =< \{c\}, \{g\} >,$$

$$e_5 =< \{c\}, \{f\} >, \ e_6 =< \{d\}, \{f\} >, \ e_7 =< \{e\}, \{f\} > .$$

The group $g_1 \subseteq U$, where

$$g_1 = \{e_1, e_2, e_3, e_4\},$$

can be represented (described) by an ordered collection of two sets of values of attributes a_1 and a_2 in the following way

$$G_{g_1} =< A_{1,t(1,g_1)}, A_{2,t(2,g_1)} >=< \{a, b, c\}, \ \{g, h\} >,$$

while the group $g_2 \subseteq U$, where

$$g_2 = \{e_5, e_6, e_7\},$$

can be represented as follows

$$G_{g_2} =< A_{1,t(1,g_2)}, A_{2,t(2,g_2)} >=< \{c, d, e\}, \ \{f\} >,$$

see Fig. 3.2.

3.5.2 *Grouping of Objects*

We want to split the set of objects U into non-empty, disjoint sets (called groups). It is required that objects in each group be *similar* in some sense, and the objects from different groups be *dissimilar*. A set of groups on U is denoted by $G(U)$.

The proposed algorithm belongs to a family of hierarchical and agglomerative clustering algorithms, so that the procedure starts with N objects as individual groups and progresses up to considering the whole set U as one group. A pair of groups corresponding to the lowest value of the measure of perturbation is coupled and a new group is formed - thereby the number of groups is decreased by one. The progress of the procedure is stopped when a fixed number of group $C \leq N$ is found. In the proposed method a measure of group perturbation is used as introduced in Definition 3.8. Additionally, it is assumed that if a certain object belongs to a defined group, then the same object must not belong to another group.

Assume that: U - set of objects, $card(U) = N$, K - number of attributes, C - predefined final number of groups. The proposed algorithm is formulated as follows:

Step 1
Iteration $c := 0$. Each of N objects constitutes one-element group in the initial set of groups $G(U)$, that is $card(G(U)) = N$ and $G(U) = \{G_{g_1}\ G_{g_2}, ..., G_{g_N}\}$.

Step 2
Iteration $c := c + 1$. Create a matrix $MPer$ of the measures of perturbations of the groups, $MPer[i, j] := Per_{OS}(G_{g_i} \mapsto G_{g_j})$, according to (3.24), for $i = 1, 2, ..., card(G(U))$, $j = 1, 2, ..., card(G(U))$, and $i \neq j$.

Step 3
Find two groups $G_{g_{i*}}$ and $G_{g_{j*}}$ that minimize the measure of perturbations of the groups.

Step 4
Create a new group in the set of groups $G(U)$ (i.e., $G_{g_{i*,j*}} := G_{g_{i*}} \oplus G_{g_{j*}}$) according to (3.18). The groups $G_{g_{i*}}$ and $G_{g_{j*}}$ are removed from the set $G(U)$, that is $card(G(U)) := card(G(U)) - 1$.

Step 5
If the required number $card(G(U)) = C$ is reached, then STOP; otherwise, return to *Step* 2 and modify the matrix $MPer$. The modification of $MPer[i, j]$ relies on removing of the $i*$-th and $j*$-th rows as well as the $i*$-th and $j*$-th columns and at the end adding a new row and a column. The new row and new column are related to the new group $G_{g_{i*,j*}}$. The disjoint set of groups $G(U)$ where $G(U) = C$, determines the resulting set of groups with assumed cardinality.

The developed approach will be illustrated by the following example. Let us consider six objects

$$U = \{e_1, e_2, e_3, e_4, e_5, e_6\}$$

described by five attributes

$$\{a_1, a_2, a_3, a_4, a_5\},$$

where attributes have the following domains

$$V_{a_1} = \{b, c, d\}, \;\; V_{a_2} = \{a, b\}, \;\; V_{a_3} = \{a, b, c\},$$

$$V_{a_4} = \{a, b\}, \;\; V_{a_5} = \{a, b, c\}.$$

For example, the objects are represented, according to (3.22), as follows:

$$Ge_1 =< \{c\}, \{b\}, \{a\}, \{a\}, \{b\} >,$$

$$Ge_2 =< \{b\}, \{a\}, \{b\}, \{a\}, \{c\} >,$$

$$Ge_3 =< \{d\}, \{b\}, \{c\}, \{a\}, \{b\} >,$$

$$Ge_4 =< \{d\}, \{a\}, \{a\}, \{b\}, \{a\} >,$$

$$Ge_5 =< \{b\}, \{a\}, \{b\}, \{b\}, \{a\} >,$$

$$Ge_6 =< \{d\}, \{b\}, \{c\}, \{a\}, \{b\} > .$$

The aim is to group the objects into two groups (i.e., $C = 2$).

At the beginning we assume, that each of 6 objects constitutes a one-element group in the initial set of groups $G(U)$. So, we have the following initial set of groups

$$G(U) =< Gg_1, Gg_2, Gg_3, Gg_4, Gg_5, Gg_6 > .$$

At first, according to (3.24), we will calculate the values of measures of perturbations (in the form the matrix $MPer$, see Table 3.1), where

$$MPer[i, j] := Per_{OS}(G_{g_i} \mapsto G_{g_j}),$$

for $i = 1, 2, ..., 6$ and $j = 1, 2, ..., 6$, for $i \neq j$. The arithmetic average was used as the aggregation operation.

Table 3.1 The measures of perturbation, iteration 1

	G_{g_1}	G_{g_2}	G_{g_3}	G_{g_4}	G_{g_5}	G_{g_6}
G_{g_1}	–	3/10	2/15	1/3	2/5	2/15
G_{g_2}	3/10	–	3/10	3/10	1/6	3/10
G_{g_3}	2/15	3/10	–	1/3	2/5	***0.0***
G_{g_4}	1/3	3/10	1/3	–	2/15	1/3
G_{g_5}	2/5	1/6	2/5	2/15	–	2/5
G_{g_6}	2/15	3/10	***0.0***	1/3	2/5	–

Table 3.2 The measures of perturbation, iteration 2

	G_{g_1}	G_{g_2}	G_{g_4}	G_{g_5}	$G_{g_{3,6}}$
G_{g_1}	–	3/10	1/3	2/5	**2/15**
G_{g_2}	3/10	–	3/10	1/6	**3/10**
G_{g_4}	1/3	3/10	–	2/15	**1/3**
G_{g_5}	2/5	1/6	2/15	–	**2/5**
$G_{g_{3,6}}$	**2/15**	**3/10**	**1/3**	**2/5**	–

Table 3.3 The measures of perturbation, iteration 3

	G_{g_2}	$G_{g_{4,5}}$	$G_{g_{1,3,6}}$
G_{g_2}	–	1/6	3/10
$G_{g_{4,5}}$	***7/30***	–	2/5
$G_{g_{1,3,6}}$	2/5	2/5	–

In Table 3.1 the minimal values appear for groups G_{g_3} and G_{g_6} (shown in bold), so from the pair of groups G_{g_3} and G_{g_6} a new group $G_{g_{3,6}}$ is created, with objects $\{e_3, e_6\}$, described as

$$Gg_{3,6} =< \{d\}, \{b\}, \{c\}, \{a\}, \{b\} > .$$

Next, the groups G_{g_3} and G_{g_6} are removed from the set $G(U)$. Thus, $card(G(U)) = 5$ and we modify the table of group perturbations $MPer$ in the way shown in Table 3.2. The newly calculated values are shown in bold.

In Table 3.2 the minimal values appear for groups G_{g_1}, $G_{g_{3,6}}$ and also for the groups G_{g_4}, G_{g_5} (i.e., 2/15). So, from the pair of groups G_{g_1} and $G_{g_{3,6}}$ a new group $G_{g_{1,3,6}}$ is created, with objects $\{e_1, e_3, e_6\}$. Likewise, from the pair of groups G_{g_4} and G_{g_5} a new group $G_{g_{4,5}}$ is created, with objects $\{e_4, e_5\}$. So, the new groups are represented as follows:

$$Gg_{4,5} =< \{b, d\}, \{a\}, \{a, b\}, \{b\}, \{a\} >,$$

$$Gg_{1,3,6} =< \{c, d\}, \{b\}, \{a, c\}, \{a\}, \{b\} > .$$

Next, the groups G_{g_1}, $G_{g_{3,6}}$ and G_{g_4}, G_{g_5} are removed from the set $G(U)$. Thus, $card(G(U)) = 3$ and we modify the table of group perturbations $MPer$ in the way shown in Table 3.3. The minimum value is shown in bold.

In Table 3.3 the minimal values appear for two groups $G_{g_{4,5}}$ and G_{g_2} (shown in bold), so from the pair of groups $G_{g_{4,5}}$ and G_{g_2} a new group $G_{g_{2,4,5}}$ is created, with objects $\{e_2, e_4, e_5\}$, represented as

$$Gg_{2,4,5} =< \{b, d\}, \{a\}, \{a, b\}, \{a, b\}, \{a, c\} > .$$

Now, the required number of groups has been already reached (i.e., $card(G(U)) = 2$), and we are able to obtain the following set of groups

$$G(U) = \{Gg_{2,4,5},\ Gg_{1,3,6}\},$$

where the groups are described in the following form

$$Gg_{1,3,6} =< \{c, d\}, \{b\}, \{a, c\}, \{a\}, \{b\} >,$$

$$Gg_{2,4,5} =< \{b, d\}, \{a\}, \{a, b\}, \{a, b\}, \{a, c\} > .$$

Thus, in this chapter we presented the concept of set perturbation, and then the measures of set perturbations. The aim of the presentation of such measures is the evaluation of similarities or distances between objects described via ordinary sets of nominal attributes. An illustrative example showed the potential possibilities of practical application of such measures of set perturbations. The following chapters are devoted to the extensions of this concept over other kinds of sets representing objects.

References

1. P. Jaccard, Distribution de la flore alpine dans le bassin des Dranses et dans quelques régions voisines. Bulletin de la Société Vaudoise des Sciences Naturelles **37**(140), 241–272 (1901)
2. G.R. Shi, Multivariate data analysis in palaeoecology and palaeobiogeography-a review. Palaeogeogr. Palaeoclimatol. Palaeoecol. **105**, 199–234 (1993)
3. A.E. Magurran, *Measuring Biological Diversity* (Blackwell, Oxford, 2004)
4. V.V. Cross, T.A. Sudkamp, *Similarity and Compatibility in Fuzzy Set Theory* (Physica-Verlag, Heidelberg, 2002)
5. T. Sorensen, A method of establishing groups of equal amplitude in plant sociology based on similarity of species content and its application to analyses of the vegetation on Danish commons. Kongelige Danske Videnskabernes Selskab **5**(4), 1–34 (1948)
6. L.R. Dice, Measures of the amount of ecologic association between species. Ecology **26**(3), 297–302 (1945)
7. H. Wolda, Similarity indices, sample size and diversity. Oceologia **50**, 296–302 (1981)
8. M.K. Vijaymeena, K. Kavitha, A survey on similarity measures in text mining. Mach. Learn. Appl. Int. J. **3**(1), 19–28 (2016)
9. A. Tversky, Features of similarity. Psychol. Rev. **84**(4), 327–352 (1977)
10. M. Krawczak, G. Szkatuła, On perturbation measure of sets-Properties. J. Autom. Mobile Robot. Intell. Syst. **8**, 41–44 (2014)
11. M. Krawczak, G. Szkatuła, On asymmetric matching between sets. Inf. Sci. **312**, 89–103 (2015)
12. J. Kacprzyk, W. Pedrycz, (eds.), *Handbook of Computational Intelligence* (Springer, 2015)
13. M. Krawczak, G. Szkatuła, A new measure of groups perturbation, in *Proceedings of the 2013 Joint IFSA World Congress NAFIPS Annual Meeting, Edmonton, Canada* (2013), pp. 1291–1296
14. M. Krawczak, G. Szkatuła, On perturbation measure of clusters-application, in *ICAISC 2013, Lecture Notes in Artificial Intelligence*, vol. 7895, Part II (Springer, Berlin, 2013), pp. 176–183

Chapter 4
Matching Between Sets with Binary Coding

4.1 Preliminaries

The binary encoding can be used for representing various kinds of descriptions of objects. It is interesting that some of the machine learning algorithms were originally designed to work only with binary-valued attributes. It happens sometimes that it is even necessary to transform a nominal attribute into several binary attributes. In the simplest case, for each value of the nominal attribute a new binary variable is generated. The number of generated binary attributes can be equal to the number of possible different splits of values of attributes into two distinct subsets. New attributes can be also constructed by using various operations in a data pre-processing step, or during the learning process.

However, application of the transformed binary attributes causes a loss of the semantics of the original attributes and may significantly increase dimensionality of the set of new binary attributes.

4.2 Representation of Nominal Values by Binary Values

Here, we will present some problems related to the replacement of an original nominal-valued attribute by a binary vector. Assume that we have a finite set V of nominal, distinct values, $V = \{v_1, v_2, \ldots, v_L\}$, and a collection of sets $\{A_1, A_2, \ldots, A_S\}$, where $A_1, A_2, \ldots, A_S \subseteq V$. Then, let us consider a pair of sets $A_i, A_j \subseteq V$, for $i, j \in \{1, 2, \ldots, S\}$.

G. Szkatuła and M. Krawczak, *Bidirectional Comparison of Nominal Sets*,
Studies in Computational Intelligence 1140,
https://doi.org/10.1007/978-3-031-53096-8_4

4.2.1 Binary Encoding of Nominal Values

First, we will introduce a binary encoding of the set $A_i \subseteq V$, which is represented by the L-dimensional binary vector

$$[w_1^i, w_2^i, ..., w_L^i],$$

for $L = card(V)$, in the following way:

$$w_l^i = \begin{cases} 1 & \text{for } v_l \in A_i \\ 0 & \text{for } v_l \notin A_i \end{cases} \tag{4.1}$$

for $l = 1, 2, ..., L$

For example, for the set $V = \{a, b, c, d\}$, the subset $A_1 = \{a\}$ is represented by a binary vector $[1, 0, 0, 0]$, and the subset A_2=$\{a, b, c, d\}$ is represented by a 4-dimensional unit vector $[1, 1, 1, 1]$, that is a 4-dimensional unit vector describes the set V.

In this way, we form a new representation of the nominal subsets, which are described by binary vectors of dimension equal to the cardinality of the set V.

4.2.2 Operations on Binary Vectors

Here, we will describe the following operations, namely subtraction, summation and intersection of binary vectors.

Let us consider two subsets A_i and A_j, where $A_i, A_j \subseteq V$, represented as binary vectors

$$[w_1^i, w_2^i, , ..., w_L^i] \text{ and } [w_1^j, w_2^j, , ..., w_L^j],$$

respectively, and the following binary vector $A_k \subseteq V$,

$$A_k = [w_1^k, w_2^k, , ..., w_L^k].$$

The basic operations are described as follows:

- *the difference of sets* $A_i \setminus A_j$ is represented by the binary vector A_k, for $l = 1, 2, ..., L$, where

$$w_l^k = \begin{cases} 1 & \text{if } w_l^i = 1 \text{ and } w_l^j = 0 \\ 0 & \text{if } w_l^j = 1 \text{ } or \text{ } w_l^i = w_l^j = 0 \end{cases} \tag{4.2}$$

- *the union of sets* $A_i \cup A_j$ is represented by the binary vector A_k, for $l = 1, 2, ..., L$, where

$$w_l^k = \begin{cases} 1 & \text{if } w_l^i = 1 \text{ or } w_l^j = 1 \\ 0 & \text{if } w_l^i = w_l^j = 0 \end{cases} \tag{4.3}$$

- *the intersection of sets* $A_i \cap A_j$ is represented by the binary vector A_k, for $l = 1, 2, ..., L$, where

$$w_l^k = \begin{cases} 1 & \text{if } w_l^i = 1 \text{ and } w_l^j = 1 \\ 0 & \text{otherwise} \end{cases} \tag{4.4}$$

For simplicity, each nominal set will be represented in this chapter by a binary vector that describes it, when it does not lead to confusion.

Thus, after replacing nominal values by digital values, in order to bring closer the concepts of operations of subtraction, summation and intersection of two sets represented by binary vectors, let us consider the following illustrative example.

Example 4.1 Assume that we have the exemplary set $V = \{a, b, c, d, e, f, m\}$ and two subsets $A_1, A_2 \subseteq V$, where

$$A_1 = \{b, c, d\} \text{ and } A_2 = \{c\}.$$

According to the introduced notation, for $card(V) = 7$, we can describe the sets A_1 and A_2 as the binary vectors, namely

$$A_1 = [0, 1, 1, 1, 0, 0, 0] \text{ and } A_2 = [0, 0, 1, 0, 0, 0, 0].$$

In this way, we can calculate the following sets:

$$A_1 \setminus A_2 = [0, 1, 0, 1, 0, 0, 0],$$

$$A_2 \setminus A_1 = [0, 0, 0, 0, 0, 0, 0],$$

$$A_1 \cup A_2 = [0, 1, 1, 1, 0, 0, 0],$$

$$A_1 \cap A_2 = [0, 0, 1, 0, 0, 0, 0].$$

□

4.2.3 Notation Used

Now, let us consider two binary vectors A_i and A_j, denoted by

$$A_i = [w_1^i, w_2^i, , ..., w_L^i] \text{ and } A_j = [w_1^j, w_2^j, , ..., w_L^j],$$

where $w_l^i, w_l^j \in \{0, 1\}$, for $l \in \{1, 2, ..., L\}$. Next, let us calculate the following numbers:

$\hat{a}$ - as the number of elements equal 1 in both vectors, (i.e., $w_l^i = w_l^j = 1$);

$\hat{b}$ - as the number of elements equal 1 in vector A_i and 0 in vector A_j, (i.e., $w_l^i = 1$ and $w_l^j = 0$);

$\hat{c}$ - as the number of elements equal 0 in vector A_i and 1 in vector A_j, (i.e., $w_l^i = 0$ and $w_l^j = 1$);

$\hat{d}$ - as the number of elements equal 0 in both vectors A_i and A_j, (i.e., $w_l^i = w_l^j = 0$).

Thus, the total sum $\hat{a} + \hat{b} + \hat{c} + \hat{d}$ (i.e., the total number of elements) is always equal to the dimension of the binary vector.

There are two interesting cases of partial summations of the numbers described above. Thus, the sum $\hat{a} + \hat{d}$ represents the total number of matches between A_i and A_j; while the sum $\hat{c} + \hat{b}$ represent the total number of mismatches between A_i and A_j, as this is shown in Table 4.1.

Again, let us assume that we have a finite set V, with cardinality $card(V) = L$. Next, let us consider two nominal subsets of the set V. Equipped with procedure (4.1) we can formulate a new representation of the nominal subsets, which are described by binary vectors of dimension equal to the cardinality of the set V (i.e., L). According to the used notation the following conditions are satisfied (cf. Krawczak and Szkatuła [19]):

Table 4.1 Binary instances between A_i and A_j

	1	0
1	$\hat{a}$	$\hat{b}$
0	$\hat{c}$	$\hat{d}$

$$\hat{a} = card(A_i \cap A_j),$$

$$\hat{b} = card(A_i \setminus A_j),$$

$$\hat{c} = card(A_j \setminus A_i),$$

$$\hat{d} = card(V \setminus (A_i \cup A_j)).$$

4.3 Examples of Proximity Measures for the Binary Set Representations

First, let us consider the domains of each attribute, which is described by a set of binary values, and denote by $dom_{l-attribute}$, for $l = 1, 2, ..., L$. Each binary attribute contains only two possible values: 1 (positive or present) or 0 (negative or absent).

Therefore, each object e_i is represented as an ordered set A_i (vector) of values for lth attribute, $l = 1, 2, ..., L$, as follows

$$A_i = [w_1^i, w_2^i, ..., w_L^i] \tag{4.5}$$

where $w_l^i \in dom_{l-attribute}$, $l = 1, 2, ..., L$.

If there is no preference as to which value should be coded as 0 and which as 1, the binary attribute is called *symmetric*. An example of such an attribute is the binary attribute *gender* with values: *male* and *female*. Both of these values are equally valuable and carry the same weight when a proximity measure is computed.

If the values of a binary attribute are not equally important, then the binary attribute is called *asymmetric*. An example of such values are the positive or negative outcomes of *a disease test*. While you say that two people who have been tested COVID positive have something in common, you cannot say that people who have not been tested positive have something in common. The more important value is usually coded as 1 (present) and the other is coded as 0 (absent), additionally, the agreement of two 1's (present - present) is more significant than the agreement of two 0's (absent - absent).

The commonly used measures accept symmetric and asymmetric binary attributes. In order to measure the similarity or dissimilarity, the binary attributes should be specified whether they are symmetric or not.

Dissimilarity for symmetric binary values can be used as a representation for the dissimilarity of vectors A_i and A_j, which can be described in the following way [19]:

$$D^{symmetric}(A_i, A_j) = \frac{\hat{b} + \hat{c}}{\hat{a} + \hat{b} + \hat{c} + \hat{d}}. \tag{4.6}$$

Dissimilarity based on asymmetric binary values, where $\hat{d}$ is considered as unimportant and is then ignored in the computation, is usually described in the following way:

$$D^{asymmetric}(A_i, A_j) = \frac{\hat{b} + \hat{c}}{\hat{a} + \hat{b} + \hat{c}}. \tag{4.7}$$

The above definitions can be easily interpreted by considering the following example.

Example 4.2 Let us consider two binary vectors

$$A_1 = [1, 0, 0, 0, 0, 0, 0, 0, 0, 0],$$

$$A_2 = [0, 0, 0, 0, 0, 0, 1, 0, 0, 1].$$

According to expressions (4.6) and (4.7) the distances between them are different for symmetric and asymmetric binary values,

$$D^{symmetric}(A_1, A_2) = \frac{3}{10},$$

$$D^{asymmetric}(A_1, A_2) = \frac{3}{3} = 1.$$

□

When both symmetric and asymmetric binary attributes occur in the same vector then a mixed approach can be applied [24].

4.3.1 Symmetric Binary Case

One can find in the literature various forms of the distance and similarity measures. In this place we would just like to recall the following definitions of the selected measures for symmetric binary cases (cf. Cross and Sudkamp [25]; Krawczak and Szkatuła [19]):

- *Jaccard extended similarity* (simple matching similarity)

$$\hat{S}_{Jaccard}(A_i, A_j) = \frac{\hat{a} + \hat{d}}{\hat{a} + \hat{b} + \hat{c} + \hat{d}}; \tag{4.8}$$

- *Mean-Manhattan distance*

$$D_{mean-Manhattan}(A_i, A_j) = \frac{\hat{b} + \hat{c}}{\hat{a} + \hat{b} + \hat{c} + \hat{d}}; \tag{4.9}$$

- *variance distance*

$$D_{variance}(A_i, A_j) = \frac{\hat{b} + \hat{c}}{4 \cdot (\hat{a} + \hat{b} + \hat{c} + \hat{d})}; \tag{4.10}$$

- *Sokal and Michener similarity*

$$S_{Sokal-Michener}(A_i, A_j) = \frac{\hat{a} + \hat{d}}{\hat{a} + \hat{b} + \hat{c} + \hat{d}}; \tag{4.11}$$

- *Faith similarity*

$$S_{Faith}(A_i, A_j) = \frac{\hat{a} + \frac{1}{2} \cdot \hat{d}}{\hat{a} + \hat{b} + \hat{c} + \hat{d}}; \tag{4.12}$$

- *Russel and Rao similarity*

$$S_{Russel-Rao}(A_i, A_j) = \frac{\hat{a}}{\hat{a} + \hat{b} + \hat{c} + \hat{d}}; \tag{4.13}$$

- *Hamann similarity*

$$S_{Hamann}(A_i, A_j) = \frac{(\hat{a} + \hat{d}) - (\hat{b} + \hat{c})}{\hat{a} + \hat{b} + \hat{c} + \hat{d}}. \tag{4.14}$$

For this representation of sets, a new measure of impact of one binary vector on another binary vector can be introduced. The new measure is called *a measure of perturbation of one binary vector by another binary vector* and can be treated as a binary version of the perturbation measure developed in this book.

- *The measure of perturbation type 1 of set A_j by set A_i* (according to Definition 3.1) can be rewritten as follows

$$Per^1_{OS}(A_i \mapsto A_j) = \frac{\hat{b}}{\hat{a} + \hat{b} + \hat{c} + \hat{d}}. \tag{4.15}$$

- *The measure of perturbation type 1 of set A_i by set A_j* is defined analogously

$$Per^1_{OS}(A_j \mapsto A_i) = \frac{\hat{c}}{\hat{a} + \hat{b} + \hat{c} + \hat{d}}. \tag{4.16}$$

The following example shows the interesting relationship between the considered proximity measures for two binary vectors.

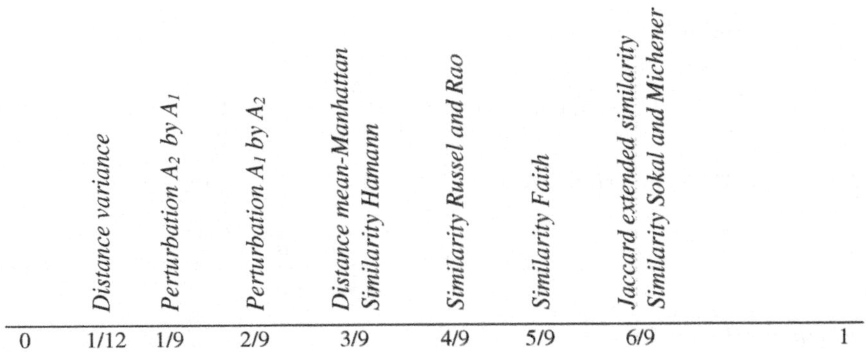

Fig. 4.1 A graphical illustration of the selected measures - symmetric binary case [19]

Example 4.3 Let us consider the following two binary vectors

$$[1, 1, 1, 1, 0, 1, 0, 0, 0] \text{ and } [1, 1, 0, 1, 1, 1, 1, 0, 0].$$

The problem is to calculate the degrees of proximity between these vectors. The values of the measures of perturbation and the measures, like Jaccard's extended similarity, mean-Manhattan distance, variance distance, Sokal and Michener similarity, Faith similarity and Russel and Rao similarity are compared. For the considered two vectors, the graphic illustration of the calculated measures and the measures of vectors perturbation is shown in Fig. 4.1.

It is easy to notice, for the considered two vectors, that different measures of set proximity generate different respective values, in general. Therefore, it seems that there is no way to indicate which measure is generally better, or it is impossible to answer the following question: whether there exists the best measure for establishing proximity between two vectors.

In Fig. 4.1, the different values of the proximity measures illustrate just the case of the considered two vectors, and any other pair of binary vectors can give different values of the proximity measures. □

The same problem exists for selecting the best measure of proximity between compared objects described by nominal attributes. Many of the known proximity measures were developed especially in order to consider the real data problems. Obviously, each of these measures provides different possibilities of selection of close or remote objects. It can happen that some measures are not able to give any rational proximity results for vector or object matching. Because of that, there are many proximity measures used in the data analysis, the proper selection depends on the nature of data and it is a heuristic process.

4.3.2 Asymmetric Binary Case

We can find in the literature (see, e.g., Choi et al. [3]), various forms of the distance and similarity measures for asymmetric binary cases. Here, we would like to recall the following selected definitions of the measures:

- *Jaccard similarity*

$$S_{Jaccard}(A_i, A_j) = \frac{\hat{a}}{\hat{a} + \hat{b} + \hat{c}}; \tag{4.17}$$

- *Dice similarity*

$$S_{Dice}(A_i, A_j) = \frac{2\hat{a}}{2\hat{a} + \hat{b} + \hat{c}}; \tag{4.18}$$

- *Nei − Li similarity*

$$S_{Nei-Li}(A_i, A_j) = \frac{2\hat{a}}{(\hat{a} + \hat{b}) + (\hat{a} + \hat{c})}; \tag{4.19}$$

- The $3W$ − *Jaccard similarity*

$$S_{3W-Jaccard}(A_i, A_j) = \frac{3\hat{a}}{3\hat{a} + \hat{b} + \hat{c}}; \tag{4.20}$$

- *Sorgenfrei similarity*

$$S_{Sorgenfrei}(A_i, A_j) = \frac{\hat{a}^2}{(\hat{a} + \hat{b})(\hat{a} + \hat{c})}; \tag{4.21}$$

- *Tanimoto similarity*

$$S_{Tanimoto}(A_i, A_j) = \frac{\hat{a}}{(\hat{a} + \hat{b}) + (\hat{a} + \hat{c}) - \hat{a}}; \tag{4.22}$$

- *Sokal – Sneath I similarity*

$$S_{Sokal-Sneath}(A_i, A_j) = \frac{\hat{a}}{\hat{a} + 2\hat{b} + 2\hat{c}}; \tag{4.23}$$

- *Driver–Kroeber similarity*

$$S_{Driver-Kroeber}(A_i, A_j) = \frac{\hat{a}}{2}\left(\frac{1}{\hat{a} + \hat{b}} + \frac{1}{\hat{a} + \hat{c}}\right); \tag{4.24}$$

- *Lance–Williams distance*

$$D_{Lance-Williams}(A_i, A_j) = \frac{\hat{b} + \hat{c}}{2\hat{a} + \hat{b} + \hat{c}}; \tag{4.25}$$

- *Bray–Curtis distance*

$$D_{Bray-Curtis}(A_i, A_j) = \frac{\hat{b} + \hat{c}}{2\hat{a} + \hat{b} + \hat{c}}. \tag{4.26}$$

- *The measure of perturbation type 2 of set A_j by set A_i* (according to Definition 3.2) can be rewritten as follows

$$Per^2_{OS}(A_i \mapsto A_j) = \frac{\hat{b}}{\hat{a} + \hat{b} + \hat{c}}. \tag{4.27}$$

- *The measure of perturbation type 2 of set A_i by set A_j* can be rewritten as follows

$$Per^2_{OS}(A_j \mapsto A_i) = \frac{\hat{c}}{\hat{a} + \hat{b} + \hat{c}}. \tag{4.28}$$

Now, let us consider the following example, which illustrates the mutual relationships between the above proximity measures.

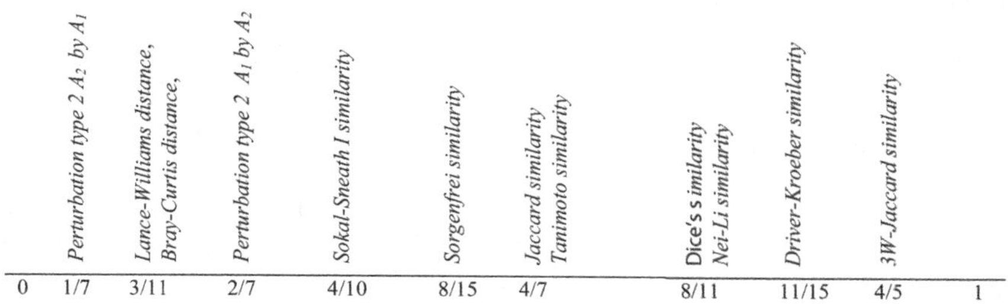

Fig. 4.2 A graphical illustration of the selected measures - asymmetric binary case [26]

Example 4.4 Let us consider two 9-dimensional binary vectors A_1 and A_2, where

$$A_1 = [1, 1, 1, 1, 0, 1, 0, 0, 0] \text{ and } A_2 = [1, 1, 0, 1, 1, 1, 1, 0, 0].$$

The problem is to calculate the degrees of proximity between these vectors. The values of the measures of perturbation of type 2 and the selected measures are compared. It seems that the best way to illustrate the proximity measure relationships is the graphic illustration shown in Fig. 4.2.

Again, it must be emphasized that the calculated measure values were obtained for these two exemplary specific binary vectors. □

It is obvious that object proximity measures are not universal and applied for the same objects return different values. In general, the measures of object proximities known in the literature are developed and designed for specified data or even for the considered data mining problems. It seems that the proposed measure of perturbation of type 2 of one binary vector by another binary vector can be considered as more general because we did not assume specify any primary conditions for the considered data set.

4.3.3 Measures of Perturbation Versus Selected Proximity Measures of Sets Coded Through Binary Representation

The different measures for binary data, recalled above, can be expressed as some functions of the measures of perturbations of type 1.

Here, according to Eqs. 4.8, 4.9, 4.10, 4.11, 4.12, 4.13 and 4.14 we can write equivalent reformulations of the selected measures, based on our idea of the set perturbation measures [19].

Corollary 4.1 *Equivalent reformulations of the selected measures can be formulated as follows:*

(1) *Extended Jaccard similarity*

$$\hat{S}_{Jaccard}(A_i, A_j) = 1 - \left(Per^1_{OS}(A_i \mapsto A_j) + Per^1_{OS}(A_j \mapsto A_i)\right); \tag{4.29}$$

(2) *Mean-Manhattan distance*

$$D_{mean-Manhattan}(A_i, A_j) = Per^1_{OS}(A_i \mapsto A_j) + Per^1_{OS}(A_j \mapsto A_i); \tag{4.30}$$

(3) *The variance distance*

$$D_{variance}(A_i, A_j) = \frac{1}{4}\left(Per^1_{OS}(A_i \mapsto A_j) + Per^1_{OS}(A_j \mapsto A_i)\right); \tag{4.31}$$

(4) *Sokal and Michener similarity*

$$S_{Sokal-Michener}(A_i, A_j) = 1 - \left(Per^1_{OS}(A_i \mapsto A_j) + Per^1_{OS}(A_j \mapsto A_i)\right); \tag{4.32}$$

(5) *Faith similarity*

$$S_{Faith}(A_i, A_j) =$$

$$= \frac{1}{2}\left(1 + \frac{card(A_j \cap A_i)}{card(V)} - (Per^1_{OS}(A_i \mapsto A_j) + Per^1_{OS}(A_j \mapsto A_i))\right); \tag{4.33}$$

(6) *Russel and Rao similarity*

$$S_{Russel-Rao}(A_i, A_j) =$$

$$= \frac{card(A_j \cup A_i)}{card(V)} - \left(Per^1_{OS}(A_i \mapsto A_j) + Per^1_{OS}(A_j \mapsto A_i)\right); \tag{4.34}$$

(7) *Hamann similarity*

$$S_{Hamann}(A_i, A_j) = 1 - 2 \cdot \left(Per^1_{OS}(A_i \mapsto A_j) + Per^1_{OS}(A_j \mapsto A_i)\right). \tag{4.35}$$

Proof According to (4.8), (4.9), (4.10), (4.11), (4.12), (4.13), (4.14) and the new representation of nominal subsets, which are described by binary vectors, the selected measures can be rewritten as follows:

The Jaccard extended similarity can be written as:

$$\hat{S}_{Jaccard}(A_i, A_j) = \frac{\hat{a} + \hat{d}}{\hat{a} + \hat{b} + \hat{c} + \hat{d}} = \frac{card(A_i \cap A_j) + card(V \setminus (A_i \cup A_j))}{card(V)} =$$

$$= \frac{card(A_i \cap A_j) + card(V) - card(A_i \cup A_j)}{card(V)} =$$

$$= 1 - \frac{card(A_i \cup A_j) - card(A_i \cap A_j)}{card(V)} = 1 - \frac{card(A_i \setminus A_j) + card(A_j \setminus A_i)}{card(V)} =$$

$$= 1 - (Per^1_{OS}(A_i \mapsto A_j) + Per^1_{OS}(A_j \mapsto A_i)).$$

The mean-Manhattan distance can be written as:

$$D_{mean-Manhattan}(A_i, A_j) = \frac{\hat{b} + \hat{c}}{\hat{a} + \hat{b} + \hat{c} + \hat{d}} =$$

$$= \frac{card(A_j \setminus A_i) + card(A_i \setminus A_j)}{card(A_i \cap A_j) + card(A_j \setminus A_i) + card(A_i \setminus A_j) + card(V \setminus (A_i \cup A_j)} =$$

$$= \frac{card(A_j \setminus A_i) + card(A_i \setminus A_j)}{card(V)} = \frac{card(A_j \setminus A_i)}{card(V)} + \frac{card(A_i \setminus A_j)}{card(V)} =$$

$$= Per^1_{OS}(A_j \mapsto A_i) + Per^1_{OS}(A_i \mapsto A_j).$$

The variance distance can be written as:

$$D_{variance}(A_i, A_j) = \frac{\hat{b} + \hat{c}}{4 \cdot (\hat{a} + \hat{b} + \hat{c} + \hat{d})} =$$

$$= \frac{card(A_j \setminus A_i) + card(A_i \setminus A_j)}{4 \cdot (card(A_i \cap A_j) + card(A_j \setminus A_i) + card(A_i \setminus A_j) + card(V \setminus (A_i \cup A_j))} =$$

$$= \frac{card(A_j \setminus A_i) + card(A_i \setminus A_j)}{4 \cdot (card(V))} =$$

$$= 1/4 \cdot (Per^1_{OS}(A_j \mapsto A_i) + Per^1_{OS}(A_i \mapsto A_j)).$$

The Sokal and Michener similarity can be written as:

$$S_{Sokal-Michener}(A_i, A_j) = \frac{\hat{a} + \hat{d}}{\hat{a} + \hat{b} + \hat{c} + \hat{d}} =$$

$$= \frac{card(A_i \cap A_j) + card(V \setminus (A_i \cup A_j))}{card(V)} =$$

$$= \frac{card(A_i \cap A_j) + card(V) - card(A_i \cup A_j)}{card(V)} =$$

$$= 1 - \frac{card(A_i \cup A_j) - card(A_i \cap A_j)}{card(V)} = 1 - \frac{card(A_i \setminus A_j) + card(A_j \setminus A_i)}{card(V)} =$$

$$= 1 - (Per^1_{OS}(A_i \mapsto A_j) + Per^1_{OS}(A_j \mapsto A_i)).$$

The Faith similarity can be written as:

$$S_{Faith}(A_i, A_j) = \frac{\hat{a} + \frac{1}{2} \cdot \hat{d}}{\hat{a} + \hat{b} + \hat{c} + \hat{d}} =$$

$$= \frac{card(A_i \cap A_j) + 1/2 \cdot card(V \setminus (A_i \cup A_j))}{card(A_i \cap A_j) + card(A_j \setminus A_i) + card(A_i \setminus A_j) + card(V \setminus (A_i \cup A_j))} =$$

$$= \frac{card(A_j \cap A_i) + card(A_j \cap A_i) + card(V \setminus (A_i \cup A_j))}{2 \cdot card(V)} =$$

$$= \frac{card(A_j \cap A_i) + card(V) - card(A_i \setminus A_j) - card(A_j \setminus A_i)}{2 \cdot card(V)} =$$

$$= \frac{card(A_j \cap A_i) + card(V)}{2 \cdot card(V)} - \frac{card(A_i \setminus A_j) + card(A_j \setminus A_i)}{2 \cdot card(V)} =$$

$$= \frac{1}{2}\left(1 + \frac{card(A_j \cap A_i)}{card(V)} - (Per^1_{OS}(A_i \mapsto A_j) + Per^1_{OS}(A_j \mapsto A_i))\right).$$

The Russel and Rao similarity can be written as:

$$S_{Russel-Rao}(A_i, A_j) = \frac{\hat{a}}{\hat{a} + \hat{b} + \hat{c} + \hat{d}} =$$

$$= \frac{card(A_i \cap A_j)}{card(A_i \cap A_j) + card(A_j \setminus A_i) + card(A_i \setminus A_j) + card(V \setminus (A_i \cup A_j))} =$$

$$= \frac{card(A_i \cup A_j) - card(A_i \setminus A_j) - card(A_j \setminus A_i)}{card(V)} =$$

$$= \frac{card(A_i \cup A_j)}{card(V)} - (Per^1_{OS}(A_i \mapsto A_j) + Per^1_{OS}(A_j \mapsto A_i)).$$

The Hamann similarity can be written as:

$$S_{Hamann}(A_i, A_j) = \frac{(\hat{a} + \hat{d}) - (\hat{b} + \hat{c})}{\hat{a} + \hat{b} + \hat{c} + \hat{d}} =$$

$$= \frac{card(A_i \cap A_j) + card(V \setminus (A_i \cup A_j)) - (card(A_j \setminus A_i) + card(A_i \setminus A_j))}{card(A_i \cap A_j) + card(A_j \setminus A_i) + card(A_i \setminus A_j) + card(V \setminus (A_i \cup A_j))} =$$

$$= \frac{card(V) - card(A_i \setminus A_j - card(A_j \setminus A_i) - card(A_j \setminus A_i) - card(A_i \setminus A_j)}{card(V)} =$$

$$= \frac{card(V) - 2 \cdot card(A_i \setminus A_j) - 2 \cdot card(A_j \setminus A_i)}{card(V)} =$$

$$= 1 - 2 \cdot (Per^1_{OS}(A_i \mapsto A_j) + Per^1_{OS}(A_j \mapsto A_i)).$$

□

According to the above considerations we can claim that the here introduced measure of perturbation of sets is quite a general measure, which can be successfully used to redefine many set similarity measures.

Additionally, we can specify an interesting property of the measures of perturbation of type 2 for the L-dimensional binary vectors and the Jaccard coefficient, presented as the following Corollary 4.2 (cf. Krawczak and Szkatuła [19]).

Corollary 4.2 *The sum of the measures of perturbation of type 2 for L-dimensional binary vectors A_i and A_j and Jaccard's coefficient satisfies the following relationship*

$$Per^2_{OS}(A_i \mapsto A_j) + Per^2_{OS}(A_j \mapsto A_i) + S_{Jaccard}(A_i, A_j) = 1. \tag{4.36}$$

Proof According to Eqs. 4.27, 4.28 and 4.17, the left side of the equation can be rewritten as follows

$$Per^2_{OS}(A_i \mapsto A_j) + Per^2_{OS}(A_j \mapsto A_i) + S_{Jaccard}(A_i, A_j) =$$

$$= \frac{\hat{b}}{\hat{a} + \hat{b} + \hat{c}} + \frac{\hat{c}}{\hat{a} + \hat{b} + \hat{c}} + \frac{\hat{a}}{\hat{a} + \hat{b} + \hat{c}} = 1.$$

□

4.4 Illustrative Examples

Let us discuss two meaningful applications of the set perturbation measures. The first application is devoted to showing how the concept of set perturbation is applied to the case of binary representation of sets. For this case the task of grouping of binary vectors has been discussed. The second application concerns the task of comparing three sets, in a binary representation, in pairs and selecting the pair that is the most similar.

4.4.1 Grouping of Binary Vectors

Let us consider a finite collection of binary vectors $\{A_1, A_2, ..., A_S\}$ of dimension L, where

$$A_i = [w_1^i, w_2^i, ..., w_L^i],$$

and where $w_l^i \in \{0, 1\}$, $l = 1, 2, ..., L$ for $i \in \{1, 2, ..., S\}$.

The task is to group "the most similar" vectors in order to obtain a new set of vectors, say, number C of vectors.

In order to solve this problem we propose a hierarchical agglomerative approach with elements of the measures of perturbation. The procedure starts with S vectors and the pair of vectors described by the lowest value of measure of perturbation is coupled and the new vector is formed - thereby the number of vectors is decreased by one. The progress of the procedure is stopped when the prescribed number C of the new vectors generated.

The operations of subtraction $A_i \setminus A_j$, summation $A_i \cup A_j$ and intersection $A_i \cap A_j$ of vectors, for $i, j \in \{1, 2, ..., S\}$ are calculated according to the formulas (4.2), (4.3) and (4.4), respectively.

The basic steps of the proposed algorithm are shown below:

Step 1

The collection ia given of S binary vectors $\{A_1, A_2, ..., A_S\}$, C is the assumed final number of vectors, iteration:= 0.

Step 2

Iteration := iteration + 1. Create the matrix D_{Per} of measures of perturbation $d_{ij} = Per^1_{OS}(A_i \mapsto A_j)$ and $d_{ji} = Per^1_{OS}(A_j \mapsto A_i)$ of the vectors (see Table 4.2).

Step 3

Find two vectors (A_{i^*}, A_{j^*}) corresponding to the minimal values of the measure of perturbation.

Table 4.2 The matrix $D_{Per} = [d_{ij}]$, $i, j = 1, 2, ...S$

	A_1	A_2	...	A_i	...	A_j	...	A_S
A_1	0	d_{12}	...	d_{1i}	...	d_{1j}	...	d_{1S}
A_2	d_{21}	0	...	d_{2i}	...	d_{2j}	...	d_{2S}
$\vdots$	$\vdots$	$\vdots$	$\ddots$	$\vdots$	$\ddots$	$\vdots$	$\ddots$	$\vdots$
A_i	d_{i1}	d_{i2}	...	0	...	d_{ij}	...	d_{iS}
$\vdots$	$\vdots$	$\vdots$	$\ddots$	$\vdots$	$\ddots$	$\vdots$	$\ddots$	$\vdots$
A_j	d_{j1}	d_{j2}	...	d_{ji}	...	0	...	d_{jS}
$\vdots$	$\vdots$	$\vdots$	$\ddots$	$\vdots$	$\ddots$	$\vdots$	$\ddots$	$\vdots$
A_S	d_{S1}	d_{S2}	...	d_{Si}	...	d_{Sj}	...	0

Step 4

Create a new binary vector $A_{i^*} \cup A_{j^*}$. The number of vectors considered is decreased by one.

Step 5

If the obtained number of vectors equals C then STOP; otherwise return to *Step* 2 and modify the matrix of measures of perturbation.

The described approach will be presented by the following illustrative example.

Example 4.5 We consider the set V of nominal values,

$$V = \{a, b, c, d, e, f, g, h, m, n, o, p\}$$

and six subsets

$$A_1 = \{a, c, d, f\},\quad A_2 = \{a, c, d, e\},\quad A_3 = \{a, g, n, o\},$$

$$A_4 = \{b, e, f, g\},\quad A_5 = \{b, h, o, p\}\quad \text{and}\quad A_6 = \{b, m, n, p\}.$$

According to (4.1), each subset $A_i, i = 1, 2, ..., 6$, can be represented by a binary vector. The exemplary set A_1 is represented by

$$[1, 0, 1, 1, 0, 1, 0, 0, 0, 0, 0, 0].$$

Thus, we have the collection of binary vectors shown in Table 4.3.

The problem is formulated as comparing vectors, and next selecting the pair of vectors giving the lowest value of perturbation, and joining this pair of vectors into one vector.

Thus, the procedure starts with 6 vectors and is stopped when two vectors are found (i.e., $C = 2$ by assumption).

In iteration 1, the values of measures of perturbation $Per^1_{OS}(A_i \mapsto A_j)$, for $i, j \in \{1, 2, ..., 6\}$, are calculated, as shown in Table 4.4.

The minimal values of the perturbation measure appear for the pair (A_1, A_2); then we create the new vector $A_1 \cup A_2$, and the number of vectors is decreased by one.

Table 4.3 The binary vectors considered

A_1	1	0	1	1	0	1	0	0	0	0	0	0
A_2	1	0	1	1	1	0	0	0	0	0	0	0
A_3	1	0	0	0	0	0	1	0	0	1	1	0
A_4	0	1	0	0	1	1	1	0	0	0	0	0
A_5	0	1	0	0	0	0	0	1	0	0	1	1
A_6	0	1	0	0	0	0	0	0	1	1	0	1

Table 4.4 The measures of perturbation, iteration 1

	A_1	A_2	A_3	A_4	A_5	A_6
A_1	–	**1/12**	3/12	3/12	4/12	4/12
A_2	**1/12**	–	3/12	3/12	4/12	4/12
A_3	3/12	3/12	–	3/12	3/12	3/12
A_4	3/12	3/12	3/12	–	3/12	3/12
A_5	4/12	4/12	3/12	3/12	–	2/12
A_6	4/12	4/12	3/12	3/12	2/12	–

Table 4.5 The binary vectors considered, iteration 3

$A_1 \cup A_2$	1	0	1	1	1	1	0	0	0	0	0	0
A_3	1	0	0	0	0	0	1	0	0	1	1	0
A_4	0	1	0	0	1	1	1	0	0	0	0	0
$A_5 \cup A_6$	0	1	0	0	0	0	0	1	1	1	1	1

Table 4.6 The measures of perturbation, iteration 3

	$A_1 \cup A_2$	A_3	A_4	$A_5 \cup A_6$
$A_1 \cup A_2$	–	4/12	3/12	5/12
A_3	3/12	–	3/12	**2/12**
A_4	**2/12**	3/12	–	3/12
$A_5 \cup A_6$	6/12	4/12	5/12	–

Table 4.7 The binary vectors after iteration 3

$A_1 \cup A_2 \cup A_4$	1	1	1	1	1	1	1	0	0	0	0	0
$A_5 \cup A_6 \cup A_3$	1	1	0	0	0	0	1	1	1	1	1	1

In the next iteration, we create the matrix of measures of perturbation for the five vectors. The minimal values of the perturbation measure appear for the pair (A_5, A_6) and the new vector $A_5 \cup A_6$ is created, as shown in Table 4.5.

Next, the procedure of calculating the perturbation measures is repeated, this time for four vectors, as shown in Table 4.6.

The minimal values of the measure perturbation appear for the pairs of vectors $(A_4,\ A_1 \cup A_2)$ and $(A_3,\ A_5 \cup A_6)$, and so the new vectors $A_1 \cup A_2 \cup A_4$ and $A_5 \cup A_6 \cup A_3$ can be created, as shown in Table 4.7.

Thereby, we obtained two final binary vectors

$$A_1 \cup A_2 \cup A_4 = [1, 1, 1, 1, 1, 1, 1, 0, 0, 0, 0, 0], \text{ and}$$

$$A_5 \cup A_6 \cup A_3 = [1, 1, 0, 0, 0, 0, 1, 1, 1, 1, 1, 1].$$

The procedure was performed for C=2. Table 4.7 can be easily interpreted, namely the first binary vector corresponds to the set of nominal values

$$A_1 \cup A_2 \cup A_4 = \{a, b, c, d, e, f, g\},$$

and the second binary vector to the set

$$A_5 \cup A_6 \cup A_3 = \{a, b, g, h, m, n, o, p\}.$$

□

4.4.2 Validations of Perturbation Measure

Here we will give an illustrative example of three sets where the task is to find out which pair of the sets gives the greatest proximity. The aim is to check how the selected measures of proximity confirm or contradict our intuition, for the selected data set.

Let us consider the set V of nominal values,

$$V = \{a, b, c, d, e, f, g, h\}$$

and three subsets

$$A_1 = \{a, b, c, d, h\}, \quad A_2 = \{a\} \text{ and } A_3 = \{c, d, e\}.$$

According to Eq. 4.1, each subset A_i, $i = 1, 2, 3$, can be represented by a binary vector $[w_1^i, w_2^i, ..., w_8^i]$ of dimension 8, where $w_l^i \in \{0, 1\}, \forall l \in \{1, 2, ..., 8\}$.

Hence, the first set A_1 is represented by the following vector

$$[1, 1, 1, 1, 0, 0, 0, 1],$$

the set A_2 by the vector

$$[1, 0, 0, 0, 0, 0, 0, 0],$$

and the set A_3 by the vector

$$[0, 0, 1, 1, 1, 0, 0, 0].$$

In order to visualize these three vectors we use a two-dimensional diagram. Each entry of the diagram represents respective values of

$$w_1^i, w_2^i, w_3^i, w_4^i \text{ (rows), and}$$

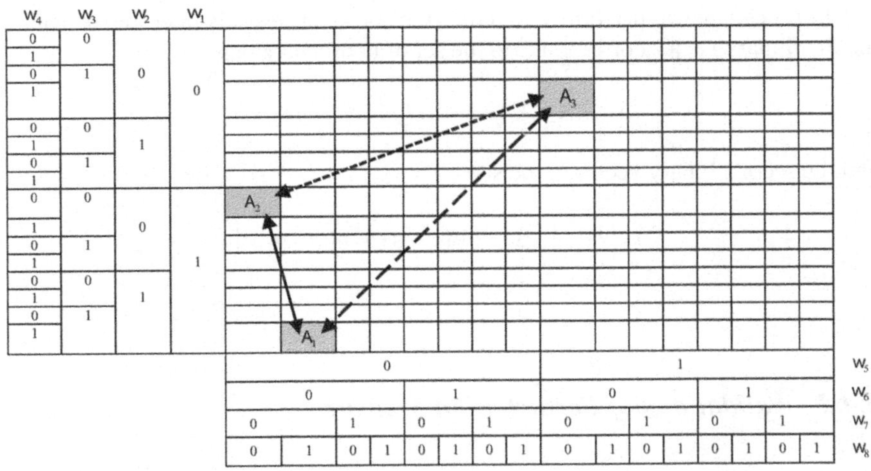

Fig. 4.3 Visualization of three considered vectors [19]

$$w_5^i, w_6^i, w_7^i, w_8^i \text{ (columns)},$$

for $i = 1, 2, 3$. In Fig. 4.3 all the vectors are shown. The arrows symbolize the distances between points A_1 and A_2; A_1 and A_3; A_2 and A_3.

In this diagram, the distance between the first pair of points (A_1, A_2) seems to be shorter than the distance between the pairs of points (A_1, A_3) or (A_2, A_3). In other words, intuitively, it seems that the pair of sets A_1 and A_2 displays the greatest proximity.

The values of the measures of perturbation (3.8), (3.9) and three other selected measures: the overlap coefficient (3.5), the Jaccard coefficient (3.1), and the Dice similarity (3.4) are calculated. The results are given in Table 4.8. The biggest values of the degree of proximity for each considered measure (i.e., the minimum values of the perturbations of sets and the maximum values of the measures of similarity), are in bold.

Table 4.8 is constructed in such a way that the symmetric or asymmetric properties of the used proximity measures are emphasized. Three of them are symmetric, while the perturbation measures are asymmetric, namely the perturbation $Per_{OS}^1(A_1 \mapsto A_2)$ is not equal to the perturbation $Per_{OS}^1(A_2 \mapsto A_1)$. However, there exists other perturbations, which are symmetric, say $Per_{OS}^1(A_2 \mapsto A_3) = Per_{OS}^1(A_3 \mapsto A_2)$.

It is easy to notice that for the applied proximity measure the results are ambiguous. The minimum value of the set perturbation appears for the pair (A_2, A_1),

$$Per_{OS}^1(A_2 \mapsto A_1) = 0.$$

Table 4.8 The values of the selected proximity measures

	A_1	A_2	A_3
A_1	–	$Per^1_{OS}(A_1 \mapsto A_2) = 1/2$ $S_{Jaccard}(A_1, A_2) = 1/5$ $S_{Dice}(A_1, A_2) = 1/3$ $S_{Ovl}(A_1, A_2) = \mathbf{1}$	$Per^1_{OS}(A_1 \mapsto A_3) = 3/8$ $S_{Jaccard}(A_1, A_3) = \mathbf{1/3}$ $S_{Dice}(A_1, A_3) = \mathbf{1/2}$ $S_{Ovl}(A_1, A_3) = 2/3$
A_2	$Per^1_{OS}(A_2 \mapsto A_1) = \mathbf{0}$ $S_{Jaccard}(A_2, A_1) = 1/5$ $S_{Dice}(A_2, A_1) = 1/3$ $S_{Ovl}(A_2, A_1) = \mathbf{1}$	–	$Per^1_{OS}(A_2 \mapsto A_3) = 1/8$ $S_{Jaccard}(A_2, A_3) = 0$ $S_{Dice}(A_2, A_3) = 0$ $S_{Ovl}(A_2, A_3) = 0$
A_3	$Per^1_{OS}(A_3 \mapsto A_1) = 1/8$ $S_{Jaccard}(A_3, A_1) = \mathbf{1/3}$ $S_{Dice}(A_3, A_1) = \mathbf{1/2}$ $S_{Ovl}(A_3, A_1) = 2/3$	$Per^1_{OS}(A_3 \mapsto A_2) = 1/8$ $S_{Jaccard}(A_3, A_2) = 0$ $S_{Dice}(A_3, A_2) = 0$ $S_{Ovl}(A_3, A_2) = 0$	–

Similarly, the maximum value of the overlap similarity appears for the pair (A_1, A_2),

$$S_{Ovl}(A_1, A_2) = 1.$$

According to these measures the sets A_1 and A_2 are located mutually more closely than the other pairs of the sets. Indeed, our intuition is confirmed only by the perturbation measure and by the overlap coefficient.

Other considered measures, namely the Jaccard coefficient and the Dice similarity indicate that it is the set A_1 and the set A_3 that are the closest to each other. The maximum value of the Jaccard similarity appears for the pair (A_1, A_3), that is

$$S_{Jaccard}(A_1, A_3) = \frac{1}{3},$$

and the maximum value of the Dice similarity appears also for the pair (A_1, A_3), that is

$$S_{Dice}(A_1, A_3) = \frac{1}{2}.$$

Based on this example, we can claim that application of the measure of set perturbation provides more information and can be considered as a competitive measure of set proximity.

References

1. F. Attneave, Dimensions of similarity. Am. J. Psychol. **63**(4), 516–556 (1950)
2. N. Goodman, Seven strictures on similarity, in *Problems and Projects* (Bobs-Merril, New York, 1972), pp. 437–450

3. S. Choi, S. Cha, C.C. Tappert, A survey of binary similarity and distance measures. Syst. Cybern. Inf. **8**(1), 43–48 (2010)
4. M. Tanca, S. Grossberg, B. Pinna, Probing perceptual antinomies with the watercolor illusion and explaining how the brain resolves them (PDF). Seeing Perceiv. **23**, 295–333 (2010)
5. R.N. Shepard, Stimulus and response generalization: a stochastic model relating generalization to distance in psychological space. Psychometrika **22**, 325–345 (1957)
6. R.N. Shepard, Stimulus and response generalization: tests of a model relating generalization to distance in psychological space. J. Exp. Psychol. **6**, 509–523 (1958)
7. W.S. Torgerson, *Theory and Methods of Scaling* (Wiley, New York, 1958)
8. W.S. Torgerson, Multidimensional scaling of similarity. Psychometrika **30**, 379–393 (1965)
9. P. Jaccard, Distribution de la flore alpine dans le bassin des Dranses et dans quelques régions voisines. Bulletin de la Société Vaudoise des Sciences Naturelles **37**(140), 241–272 (1901)
10. G.R. Shi, Multivariate data analysis in palaeoecology and palaeobiogeography - a review. Palaeogeogr. Palaeoclimatol. Palaeoecol. **105**, 199–234 (1993)
11. A.E. Magurran, *Measuring Biological Diversity* (Blackwell, Oxford, 2004)
12. H. Wolda, Similarity indices, sample size and diversity. Oceologia **50**, 296–302 (1981)
13. Z. Hubálek, Coefficients of association and similarity, based on binary (presence-absence) data: an evaluation. Biol. Rev. **57**, 669–689 (1982)
14. C. Stanfill, D. Waltz, Toward memory-based reasoning. Commun. ACM **29**, 1213–1228 (1986)
15. A. Wojna, Analogy-based reasoning in classifier construction, in *Transactions on Rough Sets IV*. Lecture Notes in Computer Science, vol. 3700, ed. J.F. Peters, A. Skowron (2005), pp. 277–374
16. A. Tversky, Features of similarity. Psychol. Rev. **84**(4), 327–352 (1977)
17. R. Beals, D.H. Krantz, A. Tversky, The foundations of multidimensional scaling. Psychol. Rev. **75**, 127–142 (1968)
18. B. Zielman, W.J. Heiser, *Spatial Representations of Asymmetric Proximities, Research Report RR-91-10* (University of Leiden, Department of Data Theory, 1991)
19. M. Krawczak, G. Szkatuła, On asymmetric matching between sets. Inf. Sci. **312**, 89–103 (2015)
20. M. Krawczak, G. Szkatuła, Geometrical interpretation of impact of one set on another set, in *Proceedings 16th International Conference, ICAISC*. LNCS, vol. 10245 (Springer, 2017), pp. 253–262
21. M. Krawczak, G. Szkatuła, Bidirectional comparison of multi-attribute qualitative objects. Inf. Sci. **436–437**, 367–387 (2018)
22. A. Tversky, Preference, belief, and similarity, in *Selected Writings by Amos Tversky*, ed. by E. Shafir (Massachusetts Institute of Technology, MIT Press, 2004)
23. J. Kacprzyk, M. Krawczak, G. Szkatuła, On bilateral matching between fuzzy set. Inf. Sci. **402**, 244–266 (2017)
24. J. Han, M. Kamber, *Data Mining: Concepts and Techniques Elsevier* (Morgan Kaufmann, San Msteo, 2006)
25. V.V. Cross, T.A. Sudkamp, *Similarity and Compatibility in Fuzzy Set Theory* (Physica-Verlag Heidelberg, 2002)
26. M. Krawczak, G. Szkatuła, On perturbation measure of sets - properties. J. Autom., Mobile Robot. Intell. Syst. **8**, 41–44 (2014)

Chapter 5
Matching Between Multisets

5.1 Preliminaries

In the classical set theory, *a set* is a collection of distinct elements. If repeating of any element is allowed, then such a set is called *a multiset* (or *a bag*). Thus, a multiset can be understood as a set of pairs, with additional information about the multiplicity of occurring elements. For instance, an exemplary description of the multiset $\{(1,a), (3,b), (2,c)\}$ is understood so that the set of three pairs is considered, wherein there is one occurrence of the element a, three occurrences of the element b, and two occurrences of the element c.

Richard Dedekind is credited with creating the concept of multisets in 1888, in the paper "Was sind und was sollen die Zahlen?". However, the term multiset was first coined by Bruijn in a private discussion with Knuth during the 1960s (see the monograph by Knuth [1]). His suggestion is now the standard terminology. The general theory of multisets can be found in the works of Blizard [2, 3]. More about relations and functions can be found in the paper [4]. The theory of the multisets, as a natural extension of the set theory, was introduced by Cerf at al. [5], Peterson [6], and Yager [7]. Surveys of the multiset theory can be found in several papers, wherein relevant operations and their properties are investigated (e.g., [8–18]).

The applications of the multiset theory can be divided into two main groups: in mathematics (especially, the combinatorial and computational aspects) and in computer science. The papers [16, 19] contain a comprehensive survey of various applications of multisets.

In many practical applications, a multi-attribute object can occur in several copies or versions. Such problems arise when, for example, some object is evaluated by several independent experts with respect to multiple criteria, or the attributes of the object were measured in different conditions, or by different methods.

The multiple-valued attributes can be processed using transformations like *averaging* or *weighting*, and so on. However, in such a case, a collection of objects may have a different structure. Therefore, new methods of describing such

G. Szkatuła and M. Krawczak, *Bidirectional Comparison of Nominal Sets*,
Studies in Computational Intelligence 1140,
https://doi.org/10.1007/978-3-031-53096-8_5

kind of objects are required. The formalism of the multiset theory allows for taking into account all of the possible combinations of attribute values simultaneously and therefore various versions of the objects can be compared. It seems to be obvious that the multisets theory gives a very convenient mathematical methodology to describe and analyze collections of multi-attribute qualitative data with repeated values of attributes.

The multiset approach to the comparison of multi-attribute objects is applicable in data mining techniques, cluster analysis, pattern recognition, decision making, and other areas.

It is worth mentioning that there are several approaches to defining the distances or similarities between multisets. A large proportion of reported definitions of metrics is due to the need of comparing objects considered in many concrete applications. Thus, for example, Manhattan distance is similar to the Penrose metric, as well as to the Minkowski metric; and the edit distance between words appears as the evolutionary distance in biology, while the one similar to the Levenshtein distance in Coding Theory and so on [20]. Developing the most adequate distance metrics in order to evaluate the proximity between objects, seems to be very important.

In general, we can distinguish two main groups of distance metrics. In the first group, each object is considered as a point in a prescribed metric space of multisets. The magnificent reviews, dedicated to distances, can be found in the papers [20–22]. In the second group, the cardinality (or the counting measure) of multisets is considered. For instance, *bag distance* is defined as a multiset metric based on the arithmetic subtraction of two multisets (having regard to the order of subtraction) [20]. Several new types of metric spaces of multisets were proposed by Petrovsky in the papers [13–15, 23]. The metrics proposed are based on the union and the symmetric difference operations on multisets. These metrics are extensions of the symmetric difference metric and the Steinhaus distance, known earlier. Other metrics can be found in the paper by Kosters and Laros [24]. In another paper, authored by Hodgetts and Hahn [25] one can find an interesting proposal of the asymmetrical transformational account of similarity of geometric patterns.

In this chapter, we consider a finite, non-empty set of objects. Each object is described by a set of attributes, and each attribute is described by nominal values, and additionally, the values of the attributes can be repeated. In other words, the multi-attribute objects with repeating qualitative values of attributes are considered. Each object is represented by a collection of multisets drawn from the sets of values of the attributes.

Thus, formalism of the theory of multisets allows for taking into account simultaneously all the combinations of attribute values and various versions of the objects. In this chapter an effective procedure for comparing such objects as well as groups of such objects is developed and discussed. The proposed concept of perturbation of one object by another is exploited, and is considered as the difference of the multisets, representing the objects. The measure of perturbation describes in the following part of the chapter remoteness between the objects, and, in general, the measure is asymmetrical. Next, we consider the measure of perturbation of one group of objects by another group of objects. Then, we generate the description of each group in the form of classification rules. The chapter ends with a practical illustration of the proposed approach for a task of grouping of text documents.

Therefore, we recall the novel *concept of perturbation of one multiset by another multiset* which constitutes a new multiset (cf. Krawczak and Szkatuła [26]). In details, it is shown that the perturbation of one multiset by another multiset is described by a difference between these two multisets, and therefore the direction of the perturbation of multisets is emphasised.

Due to normalization of the cardinality of this difference, the measure of perturbation ranges between 0 and 1, wherein 0 indicates the lowest value of perturbation, while 1 indicates the highest value of perturbation. Two types of *measure of multiset perturbation* are studied. The first is called the measure of perturbation *type 1*, where the perturbation is normalized by the arithmetic addition of the two multisets in question (cf. Krawczak and Szkatuła [10, 11]). The second is called the measure of perturbation *type 2* (cf. Krawczak and Szkatuła [12]), where the perturbation is normalized by the union of the two multisets.

Then, a description of a group of objects as an ordered collection of the multisets, as well as a concept of perturbation of one group of objects by another group of objects are provided. In this case, the perturbation is represented by the difference between the description of one group and the description of another group.

It seems that the concept of perturbation can find wide applications in solving problems based on comparison of multisets, when the "direction" of comparison has a significant meaning. For example, the methodology allows for generating classifications rules distinguishing the considered groups of objects (e.g., the textual documents, as this is shown in Sect. 5.6). These rules can be used to classify new objects to one of the prescribed group. Another example of application of this methodology is the possibility of evaluating group distances in order to solve the clustering tasks, as this is shown in the authors' paper [27].

5.2 Brief Introduction to Multisets Theory

We consider the multisets in the so-called multiplicative form [19, 28], drawn from a non-empty and finite set V of nominal, distinct elements with cardinality L, $V = \{v_1, v_2, \ldots, v_L\}$, described as follows.

Definition 5.1 (*Multiset*) The multiset S drawn from the ordinary set V can be represented by a set of ordered pairs:

$$S = \{(k_s(v), v)\}, \forall v \in V, \tag{5.1}$$

where $k_s : V \to \{0, 1, 2, ..., m\}$, and m is an integer number.

In (5.1) the function $k_s(.)$ is called *a counting function* or *a multiplicity function*, and the value of $k_s(v)$ specifies the number of occurrences of the element

$v \in V$ in the multiset S, such that no element occurs more than m times. The element, which is not included in the multiset S has its counting function equal zero.

Let us denote by $V^m(L)$ the set of the multisets drawn from the set V. The cardinality of the set V is L, and m is the number indicated in Definition 5.1. Definition 5.1 can be rewritten in the following way

$$S = \{(k_s(v_1), v_1), \ (k_s(v_2), v_2), \ ..., (k_s(v_L), v_L)\}, \tag{5.2}$$

understood so that the element $v_1 \in V$ appears $k_s(v_1)$ times in the multiset S, the element $v_2 \in V$ appears $k_s(v_2)$ times, and so on. In the case, where $k_s(v_i) = 0$, the element $v_i \in V$ is omitted.

There is a special case, namely, we say that the multiset S is empty (denoted by $\emptyset$ or $\{\}$), if and only if $\forall\, i \in \{1, 2, ..., L\}, k_s(v_i) = 0$.

5.2.1 Operations on Multisets

Let us consider two multisets, S_1 and S_2, such that $S_1, S_2 \in V^m(L)$, as follows

$$S_1 = \{(k_{s_1}(v_1), v_1), (k_{s_1}(v_2), v_2), ..., (k_{s_1}(v_L), v_L)\},$$

$$S_2 = \{(k_{s_2}(v_1), v_1), (k_{s_2}(v_2), v_2), ..., (k_{s_2}(v_L), v_L)\}.$$

The following basic operations on the multisets are well known [10–12]:

- *the union of multisets* is the multiset defined by $S_1 \cup S_2 = \{(k_{s_1 \cup s_2}(v), v) \mid k_{s_1 \cup s_2}(v) := max\{k_{s_1}(v), k_{s_2}(v)\}, v \in V\}$,
- *the intersection of multisets* is the multiset defined by $S_1 \cap S_2 = \{(k_{s_1 \cap s_2}(v), v) \mid k_{s_1 \cap s_2}(v) := min\{k_{s_1}(v), k_{s_2}(v)\}, v \in V\}$,
- *the arithmetic addition of multisets* is the multiset defined by $S_1 \oplus S_2 = \{(k_{s_1 \oplus s_2}(v), v) \mid k_{s_1 \oplus s_2}(v) := k_{s_1}(v) + k_{s_2}(v), v \in V\}$,
- *the arithmetic subtraction of multisets* is the multiset defined by $S_1 \ominus S_2 = \{(k_{s_1 \ominus s_2}(v), v) \mid k_{s_1 \ominus s_2}(v) := max\{k_{s_1}(v) - k_{s_2}(v), 0\}, v \in V\}$,
- *the symmetric difference of multisets* is the multiset defined by $S_1 \triangle S_2 = \{(k_{s_1 \triangle s_2}(v), v) \mid k_{s_1 \triangle s_2}(v) := \mid k_{s_1}(v) - k_{s_2}(v) \mid, v \in V\}$.

Now, we recall the basic notions, related to the multisets, illustrated by simple examples, namely:

- *The support* (or *the root*) *of the multiset S drawn from a set V* (denoted by S^*), is an ordinary set defined as follows:

$$S^* = \{v \in V \mid k_s(v) > 0\}.$$

For example, the support of the multiset $S = \{(3,a), (2,d), (5,f)\}$ is constituted by $S^* = \{a, d, f\}$.

- *The cardinality of the multiset S* (denoted by $card(S)$ or $\mid S \mid$) is defined as the total number of its elements, that is

$$card(S) = \sum_{v \in V} k_s(v).$$

- *The dimensionality of the multiset S* (denoted by $dim(S)$ or /S/), is defined as the total number of various elements, that is

$$dim(S) = \sum_{v \in V} \chi_S(v),$$

χ_S is the characteristic function of the set S.

For example, for the multiset $S = \{(3,a), (3,b)\}$, $card(S) = 6$ and $dim(S) = 2$.

5.2.2 Examples of Proximity Measures of Multisets

There is a rich literature on various definitions of distance between multisets. For example, we recall the following measures (e.g., [1, 22, 24]):

- *Bray-Curtis dissimilarity*

$$D_{Bray-Curtis}(S_1, S_2) = \frac{card(S_1 \triangle S_2)}{card(S_1 \oplus S_2)}, \tag{5.3}$$

which is popular in the environmental sciences,

- *Steinhaus distance*

$$D_{Steinhaus}(S_1, S_2) = \frac{card(S_1 \triangle S_2)}{card(S_1 \cup S_2)}, \tag{5.4}$$

and three commonly used distance measures, namely

- *Chebyshev distance*

$$D_{Chebyshev}(S_1, S_2) = \max_{i \in \{1,2,\ldots,L\}} \mid k_{S_1}(v_i) - k_{S_2}(v_i) \mid, \tag{5.5}$$

- *Manhattan distance*

$$D_{Manhattan}(S_1, S_2) = \sum_{i=1}^{L} \mid k_{S_1}(v_i) - k_{S_2}(v_i) \mid, \tag{5.6}$$

- *Euclidean distance*

$$D_{Euclidean}(S_1, S_2) = \sqrt{\sum_{i=1}^{L}(k_{S_1}(v_i) - k_{S_2}(v_i))^2}. \quad (5.7)$$

Equipped with the basics of multisets, we can define the new asymmetric measure of proximity between two multisets in the forthcoming section.

5.3 Basics of Perturbation of Multisets

Comparison of the first multiset S_1 with the second multiset S_2 is understood in such a manner that the second multiset is perturbed by the first multiset to a definite degree. On the other hand, comparison of the second multiset S_2 with S_1 means that the first multiset is perturbed by the second one to another degree. It is important to notice that the direction of the perturbation is important.

5.3.1 The Concept of Perturbation of Multisets

In the papers of Krawczak and Szkatuła [10–12], *the concept of perturbation* of one multiset S_2 by another multiset S_1, denoted by $(S_1 \mapsto S_2)$, was developed, which is interpreted as a difference between one multiset and another multiset, $S_1 \ominus S_2$, in the following way:

$$(S_1 \mapsto S_2) = \{(k_{S_1 \mapsto S_2}(v), v) \mid k_{S_1 \mapsto S_2}(v) := max\{k_{S_1}(v) - k_{S_2}(v), 0\}\}. \quad (5.8)$$

The counterpart definition is analogous

$$(S_2 \mapsto S_1) = \{(k_{S_2 \mapsto S_1}(v), v) \mid k_{S_2 \mapsto S_1}(v) := max\{k_{S_2}(v) - k_{S_1}(v), 0\}\}. \quad (5.9)$$

Now, we will present the interpretation of the perturbation of one multiset by another multiset in the following example.

Example 5.8 Consider the set $V = \{a, b, c, d, e\}$ and two exemplary multisets

$$S_1 = \{(1, a), (1, e)\} \text{ and } S_2 = \{(1, a), (1, d), (3, e)\},$$

where $S_1, S_2 \in V^3(5)$. The perturbation of the multiset S_2 by the multiset S_1 is the empty multiset, because $(S_1 \mapsto S_2) = \emptyset$. The perturbation of the multiset S_1 by the multiset S_2 is the following multiset

$$(S_2 \mapsto S_1) = \{(1, d), (2, e)\}. \qquad \square$$

Note that each finite multiset drawn from the ordinary set of L elements can be shown as a point in two-dimensional space. For example, assuming that $L = 2$, the multiset $\{b, a, b, b\}$ can be written in a simplified form as $\{(1, a), (3, b)\}$ (since the order of elements is irrelevant) and by omitting the names of the elements, we get the point (1,3) in two-dimensional space.

Thus, we can consider the geometrical interpretation of the proposed concept of the perturbation in two-dimensional space, for $L = 2$.

Example 5.9 Let us assume that $card(V) = 2$ (i.e., $V = \{v_1, v_2\}$), and then consider two multisets $S_1, S_2 \in V^m(2)$,

$$S_1 = \{(k_{S_1}(v_1), v_1),\ (k_{S_1}(v_2), v_2)\},$$

$$S_2 = \{(k_{S_2}(v_1), v_1),\ (k_{S_2}(v_2), v_2)\}.$$

Each considered multiset can be represented as a point in two-dimensional space, see Fig. 5.1, and these two points have the following coordinates

$$\left(k_{S_1}(v_1), k_{S_1}(v_2)\right) \text{ and } \left(k_{S_2}(v_1), k_{S_2}(v_2)\right),$$

respectively. According to (5.8) and (5.9), the perturbation of an arbitrary multiset S_2 by another multiset S_1 is interpreted as a new multiset described as follows [10–12]:

$$(S_1 \mapsto S_2) = \{(k_{S_1 \mapsto S_2}(v_1), v_1),\ (k_{S_1 \mapsto S_2}(v_2), v_2)\} =$$

$$= \{(max\{k_{S_1}(v_1) - k_{S_2}(v_1), 0\}, v_1),\ (max\{k_{S_1}(v_2) - k_{S_2}(v_2), 0\}, v_2)\}.$$

And, in the opposite case, the perturbation of the multiset S_1 by the multiset S_2 has an analogous definition [10–12]

$$(S_2 \mapsto S_1) = \{(k_{S_2 \mapsto S_1}(v_1), v_1),\ (k_{S_2 \mapsto S_1}(v_2), v_2)\} =$$

$$= \{(max\{k_{S_2}(v_1) - k_{S_1}(v_1), 0\}, v_1),\ (max\{k_{S_2}(v_2) - k_{S_1}(v_2), 0\}, v_2)\}.$$

The two-dimensional geometrical interpretations of non-zero values of counting functions of the perturbations for the exemplary multisets S_1 and S_2 are presented in Fig. 5.1. In the figure, two perturbations are indicated (i.e., the perturbation $(S_1 \mapsto S_2)$ in the left figure, and the perturbation $(S_2 \mapsto S_1)$ in the right figure). The arrows indicate the directions of the perturbations.

When analyzing Fig. 5.1, it is easy to notice that for the exemplary multisets S_1, S_2 the perturbation of one multiset by another one creates a new multiset, obtained from these two multisets. Thus, the multiset perturbation corresponds to the difference

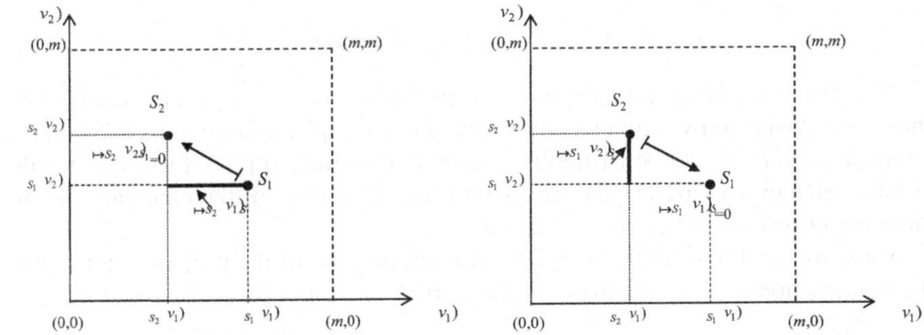

Fig. 5.1 The graphical illustrations of perturbations of the multisets S_1 and S_2

between multisets, and therefore the direction of the perturbation can be important. The following conditions

$$k_{S_1 \mapsto S_2}(v_1) = k_{S_1}(v_1) - k_{S_2}(v_1) \text{ and } k_{S_1 \mapsto S_2}(v_2) = 0,$$

as well as

$$k_{S_2 \mapsto S_1}(v_1) = 0 \text{ and } k_{S_2 \mapsto S_1}(v_2) = k_{S_2}(v_2) - k_{S_1}(v_2),$$

are satisfied. The segments marked by the thick lines in Fig. 5.1 indicate positive values of the counting functions $k_{S_1 \mapsto S_2}(v_1)$ and $k_{S_2 \mapsto S_1}(v_2)$, respectively.

In the case of the first perturbation $(S_1 \mapsto S_2)$ (left part of Fig. 5.1), the beginning of the segment is at the point

$$\big(k_{S_2}(v_1), k_{S_1}(v_2)\big),$$

and the end of the segment is at the point

$$\big(k_{S_1}(v_1), k_{S_1}(v_2)\big).$$

On the other hand, for the opposite perturbation $(S_2 \mapsto S_1)$ (right part of Fig. 5.1), the beginning of the segment is at the point $(k_{S_2}(v_1), k_{S_1}(v_2))$, and the end is at the point $(k_{S_2}(v_1), k_{S_2}(v_2))$.

Thus, the first perturbation, shown on the left side of Fig. 5.1, can be rewritten in the following multiset form

$$(S_1 \mapsto S_2) = \{(k_{S_1 \mapsto S_2}(v_1), v_1), (k_{S_1 \mapsto S_2}(v_2), v_2)\} =$$

$$= \{(k_{S_1}(v_1) - k_{S_2}(v_1), v_1), (0, v_2)\}$$

while the second perturbation, shown on the right side of Fig. 5.1, can be rewritten as

$$(S_2 \mapsto S_1) = \{(k_{S_2 \mapsto S_1}(v_1), v_1), (k_{S_2 \mapsto S_1}(v_2), v_2)\} =$$

$$= \{(0, v_1), (k_{S_2}(v_2) - k_{S_1}(v_2), v_2)\}.$$

□

Next, we will present the details of the measure of perturbation of one multiset by another multiset.

5.3.2 *Measure of Perturbation of Multisets*

As previously, let us consider two multisets $S_1, S_2 \in V^m(L)$, for the set $V = \{v_1, v_2, \ldots, v_L\}$, where m is an integer number indicated in Definition 5.1. The perturbation of one multiset by another constitutes a new multiset, and there is a problem of determining the numerical values of the multiset perturbations. For this purpose, we will present two proposals of defining the measure of perturbation of one multiset by another multiset, whose values range between 0 and 1. Value 0 indicates the lowest value of the perturbation measure, while 1 is the highest value. The definitions are based on cardinalities of the considered multisets as the functions that assign a non-negative real number to each finite multiset $S \in V^m(L)$ (i.e., $card(S) = \sum_{v \in V} k_S(v)$). At the beginning, the arithmetic subtraction of two multisets is determined and its cardinality is described, and then the result is normalized.

We distinguish two kinds of multiset perturbation measures, first, *the measure of perturbation type 1* of one multiset by another with normalization using the arithmetic addition of these two multisets $S_1 \oplus S_2$, and another *measure of perturbation type 2* with normalization based on the union of two considered multisets, $S_1 \cup S_2$.

First, let us consider the measure of the multiset perturbation type *1* of the multiset S_2 by the multiset S_1 (cf. Krawczak and Szkatuła [10, 11]).

Definition 5.2 (*Measure of perturbation type 1*) The measure of perturbation type 1 of the multiset S_2 by the multiset S_1 is defined by the mapping $Per^1_{MS} : V^m(L) \times V^m(L) \to [0, 1]$, in the following manner:

$$Per^1_{MS}(S_1 \mapsto S_2) = \frac{card(S_1 \ominus S_2)}{card(S_1 \oplus S_2)} = \frac{\sum_{i=1}^{L}(k_{S_1}(v_i) - k_{S_1 \cap S_2}(v_i))}{\sum_{i=1}^{L}(k_{S_1}(v_i) + k_{S_2}(v_i))}. \tag{5.10}$$

The intuitive meaning of the above definition can be given as follows, namely, the measure of perturbation of one multiset by another is understood as the total number of elements appearing in the multiset, which is the result of the arithmetic subtraction of these multisets. The measure is normalized by the total number of elements in the multiset formed by arithmetic addition of these multisets. The normalization causes that the measure does not exceed the value of one.

In the counterpart case, the measure of perturbation of the multiset S_1 by the multiset S_2 is defined in an analogous way:

$$Per^1_{MS}(S_2 \mapsto S_1) = \frac{card(S_2 \ominus S_1)}{card(S_2 \oplus S_1)} = \frac{\sum_{i=1}^{L}(k_{S_2}(v_i) - k_{S_2 \cap S_1}(v_i))}{\sum_{i=1}^{L}(k_{S_2}(v_i) + k_{S_1}(v_i))}. \quad (5.11)$$

Although the definitions of these two cases are analogous, there is a difference, resulting from the directional character of the arithmetic subtractions.

The measure of the perturbation type 2 is defined in the following way (cf. Krawczak and Szkatuła [12]).

Definition 5.3 (*Measure of perturbation type 2*) The measure of perturbation type 2 of the multiset S_2 by the multiset S_1 is defined by the mapping $Per^2_{MS} : V^m(L) \times V^m(L) \to [0, 1]$, in the following manner:

$$Per^2_{MS}(S_1 \mapsto S_2) = \frac{card(S_1 \ominus S_2)}{card(S_1 \cup S_2)} = \frac{\sum_{i=1}^{L}(k_{S_1}(v_i) - k_{S_1 \cap S_2}(v_i))}{\sum_{i=1}^{L} max\{k_{S_1}(v_i), k_{S_2}(v_i)\}}. \quad (5.12)$$

The definition of the counterpart case is analogous:

$$Per^2_{MS}(S_2 \mapsto S_1) = \frac{card(S_2 \ominus S_1)}{card(S_2 \cup S_1)} = \frac{\sum_{i=1}^{L}(k_{S_2}(v_i) - k_{S_2 \cap S_1}(v_i))}{\sum_{i=1}^{L} max\{k_{S_2}(v_i), k_{S_1}(v_i)\}}. \quad (5.13)$$

The measure of perturbation type 1 of multisets differs from the measure of perturbation type 2 by the different form of the denominators. Namely, in Definition 5.2 we have the arithmetic addition $S_1 \oplus S_2$, while in Definition 5.3 the union of multisets $S_1 \cup S_2$.

5.3.3 Properties of Perturbation Measure of Multisets

The measure of multiset perturbation type 1 satisfies the following properties:

Corollary 5.10 *The measure of perturbation type 1 of the multiset S_2 by the multiset S_1 satisfies the following conditions, for I= {1, 2, ...,L},*

(1) $0 \leq Per^1_{MS}(S_1 \mapsto S_2) \leq 1$
(2) $Per^1_{MS}(S_1 \mapsto S_2) = 0 \; if \; and \; only \; if \; k_{S_1}(v_i) = k_{S_1 \cap S_2}(v_i), \; \forall i \in I$
(3) $If \quad \forall i \in I, \quad k_{S_2}(v_i) = 0 \; and \quad \exists k_{S_1}(v_i) > 0, \quad for \quad i \in I, \quad then$ $Per^1_{MS}(S_1 \mapsto S_2) = 1.$

Proof By Definition 5.2.

□

The measure of perturbation type 2 of one multiset by another satisfies the following properties:

Corollary 5.11 *The measure of perturbation type 2 of the multiset S_2 by the multiset S_1 satisfies the following conditions, for I= {1, 2, ...,L},*

(1) $0 \leq Per^2_{MS}(S_1 \mapsto S_2) \leq 1$
(2) $Per^2_{MS}(S_1 \mapsto S_2) = 0 \; if \; and \; only \; if \; k_{S_1}(v_i) = k_{S_1 \cap S_2}(v_i), \forall i \in I$
(3) $If \quad \forall i \in I, k_{S_2}(v_i) = 0 \quad and \quad \exists k_{S_1}(v_i) > 0, \quad for \quad i \in I, \quad then$ $Per^2_{MS}(S_1 \mapsto S_2) = 1.$

Proof By Definition 5.3.

□

The idea of multiset perturbation can be now illustrated by the following example.

Example 5.10 Consider the following set $V = \{a, b, d, e\}$ and two exemplary multisets

$$S_1 = \{(1, a), (1, e)\} \text{ and } S_2 = \{(1, a), (1, d), (3, e)\},$$

where $S_1, S_2 \in V^4(4)$. Due to Definition 5.2, the measures of perturbation type 1 is calculated in the following way:

$$Per^1_{MS}(S_1 \mapsto S_2) = \frac{\sum_{i=1}^{4}(k_{S_1}(v_i) - k_{S_1 \cap S_2}(v_i))}{\sum_{i=1}^{4}(k_{S_1}(v_i) + k_{S_2}(v_i))} = 0,$$

$$Per^1_{MS}(S_2 \mapsto S_1) = \frac{\sum_{i=1}^{4}(k_{S_2}(v_i) - k_{S_2 \cap S_1}(v_i))}{\sum_{i=1}^{4}(k_{S_2}(v_i) + k_{S_1}(v_i))} = \frac{3}{7}.$$

□

5.3.4 Graphical Illustration of the Multiset Perturbation Measure

In order to demonstrate the meaning of the measures of perturbation, concerning both *type 1* and *type 2*, of one multiset S_2 by another multiset S_1, that is

$$Per^1_{MS}(S_1 \mapsto S_2) \text{ and } Per^2_{MS}(S_1 \mapsto S_2),$$

as well as the counterpart cases, that is

$$Per^1_{MS}(S_2 \mapsto S_1) \text{ and } Per^2_{MS}(S_2 \mapsto S_1),$$

we provide some geometrical interpretations of the measures of perturbation of the multisets in two-dimensional space and in three-dimensional space.

Example 5.11 (*Two-dimensional case, 2D*) Let us assume that $V = \{a\}$ (i.e., $L = card(V) = 1$), and consider two multisets

$$S_1 = \{(k_{S_1}(a), a)\} \text{ and } S_2 = \{(k_{S_2}(a), a)\},$$

for $S_1, S_2 \in V^5(1)$. According to Eqs. 5.10 and 5.11, the measures of perturbation type 1 have the following forms:

$$Per^1_{MS}(S_1 \mapsto S_2) = \frac{k_{S_1}(a) - k_{S_1 \cap S_2}(a)}{k_{S_1}(a) + k_{S_2}(a)},$$

$$Per^1_{MS}(S_2 \mapsto S_1) = \frac{k_{S_2}(a) - k_{S_2 \cap S_1}(a)}{k_{S_2}(a) + k_{S_1}(a)},$$

and according to Eqs. 5.12 and 5.13, the measures of perturbation type 2 have the following forms

$$Per^2_{MS}(S_1 \mapsto S_2) = \frac{k_{S_1}(a) - k_{S_1 \cap S_2}(a)}{max\{k_{S_1}(a), k_{S_2}(a)\}},$$

$$Per^2_{MS}(S_2 \mapsto S_1) = \frac{k_{S_2}(a) - k_{S_2 \cap S_1}(a)}{max\{k_{S_2}(a), k_{S_1}(a)\}}.$$

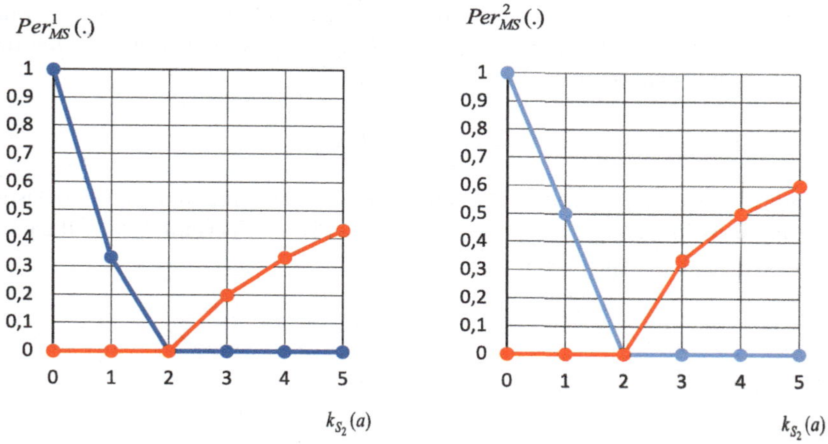

Fig. 5.2 The measures of perturbations $Per^1_{MS}(.)$ and $Per^2_{MS}(.)$: the perturbation $(S_1 \mapsto S_2)$ – the blue lines, the perturbation $(S_2 \mapsto S_1)$ – the red lines. The value of $k_{S_1}(a)$ is equal 2, and $k_{S_2}(a)$ is changing from 0 to 5

Additionally, it is assumed that the counting function for the multiset S_1 equals 2, that is

$$k_{S_1}(a) = 2,$$

while the counting function for the multiset S_2 is being changed from 0 to 5, that is, we deal with the cases of

$$k_{S_2}(a) = 0, 1, 2, 3, 4, 5.$$

In this way, we consider the pairs of the multisets: S_1 and S_2, where the multiset S_1 is fixed, that is

$$S_1 = \{(2, a)\},$$

and the second multiset S_2 is being changed as follows:

$$S_2 = \{(0, a)\}, \quad S_2 = \{(1, a)\}, \quad S_2 = \{(2, a)\},$$

$$S_2 = \{(3, a)\}, \quad S_2 = \{(4, a)\}, \quad S_2 = \{(5, a)\}.$$

Figure 5.2 shows the comparisons between the values of measures of the perturbations for such pairs of multisets S_1 and S_2.

In the left figure, the measures are displayed of the perturbation type 1, denoted by $Per^1_{MS}(.)$, while in the right-hand figure the values are displayed of the measures of the perturbation type 2, denoted by $Per^2_{MS}(.)$, for the pairs of multisets S_1 and S_2.

The figures display the changes of values of the perturbation measures with respect to the values $k_{S_2}(a)$ (which are changed from 0 to 5), for the fixed value of the function $k_{S_1}(a) = 2$.

For the first case of the perturbation (i.e., $(S_1 \mapsto S_2)$), the measures $Per^1_{MS}(S_1 \mapsto S_2)$ and $Per^2_{MS}(S_1 \mapsto S_2)$ (indicated as the points on the blue lines in Fig. 5.2) are equal 0 for

$$k_{S_1}(a) = 2 \leq k_{S_2}(a) \leq 5.$$

For the second case of the perturbation (i.e., $(S_2 \mapsto S_1)$), the values of the measures of the perturbation: $Per^1_{MS}(S_2 \mapsto S_1)$ and $Per^2_{MS}(S_2 \mapsto S_1)$ (indicated as the points on the red lines) are equal 0 for

$$0 \leq k_{S_2}(a) \leq k_{S_1}(a) = 2.$$

It is interesting to note that both curves are convex.

□

Example 5.12 (*Three-dimensional case, 3D*) Now, let us consider a case characterized by the set $V = \{a, b\}$, for which card(V)=2, and two exemplary multisets $S_1, S_2 \in V^4(2)$,

$$S_1 = \{(k_{S_1}(a), a),\ (k_{S_1}(b), b)\},$$

$$S_2 = \{(k_{S_2}(a), a),\ (k_{S_2}(b), b)\}.$$

It is assumed, additionally, that the value of each counting function for S_1 is equal 2, that is

$$k_{S_1}(a) = 2 \text{ and } k_{S_1}(b) = 2,$$

while the values of the counting function for S_2 range between 0 and 4, that is

$$k_{S_2}(a), k_{S_2}(a) \in \{0, 1, 2, 3, 4\}.$$

In this way, we consider the pairs of the multisets S_1 and S_2, where the multiset S_1 is fixed, that is

$$S_1 = \{(2, a), (2, b)\}$$

and the second multiset S_2 is being changed as follows

$$S_2 = \{(0, a), (0, b)\},\ \ S_2 = \{(0, a), (1, b)\},\ \ S_2 = \{(0, a), (2, b)\},$$

$$S_2 = \{(0, a), (3, b)\},\ \ S_2 = \{(0, a), (4, b)\},$$

$$S_2 = \{(1, a), (0, b)\},\ \ S_2 = \{(1, a), (1, b)\},\ \ S_2 = \{(1, a), (2, b)\},$$

$$S_2 = \{(1, a), (3, b)\},\ \ S_2 = \{(1, a), (4, b)\},$$

$$\ldots$$

$$S_2 = \{(4, a), (0, b)\}, \quad S_2 = \{(4, a), (1, b)\}, \quad S_2 = \{(4, a), (2, b)\},$$

$$S_2 = \{(4, a), (3, b)\}, \quad S_2 = \{(4, a), (4, b)\}.$$

Let us consider the measure of perturbation type 2 for the multisets S_1 and S_2, denoted by $Per^2_{MS}(S_2 \mapsto S_1)$, and described by Eq. 5.13:

$$Per^2_{MS}(S_2 \mapsto S_1) = \frac{\sum_{i=1}^{2}(k_{S_2}(v_i) - k_{S_2 \cap S_1}(v_i))}{\sum_{i=1}^{2} max\{k_{S_1}(v_i), k_{S_2}(v_i)\}} =$$

$$= \frac{k_{S_2}(a) + k_{S_2}(b) - k_{S_2 \cap S_1}(a) - k_{S_2 \cap S_1}(b)}{max\{k_{S_1}(a), k_{S_2}(a)\} + max\{k_{S_1}(b), k_{S_2}(b)\}}.$$

Thus, let us consider measure of perturbation type 2, for the fixed multiset

$$S_1 = \{(2, a), (2, b)\},$$

and for the changing multiset

$$S_2 = \{(k_{S_2}(a), a), \ (k_{S_2}(b), b)\},$$

(i.e., for the changing values of $k_{S_2}(a)$ and $k_{S_2}(b)$ from 0 to 4).

In the three-dimensional space, each the point has the following coordinates

$$(k_{S_2}(a), \ k_{S_2}(b), \ Per^2_{MS}(S_2 \mapsto S_1)),$$

and can be represented as a point on a plane in Fig. 5.3.

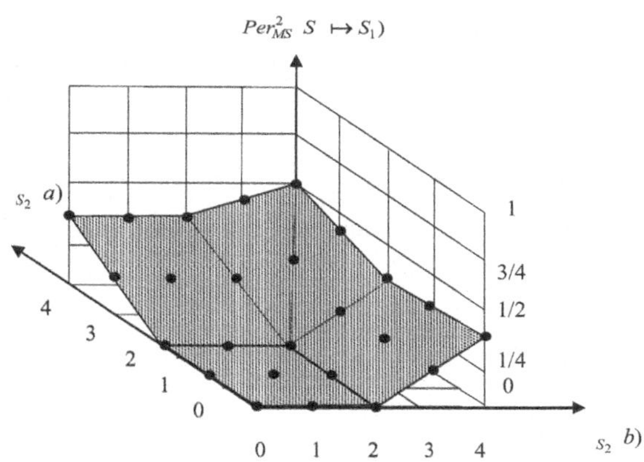

Fig. 5.3 The exemplary measures of the perturbations in the three-dimensional space

Figure 5.3 shows that the measure of the perturbation type 2, denoted by $Per^2_{MS}(S_2 \mapsto S_1)$, is equal 0 if $k_{S_2}(a) \in \{0, 1, 2\}$ and $k_{S_2}(b) \in \{0, 1, 2\}$. The value of the measure of perturbation is greater than zero if $k_{S_2}(a) \in \{3, 4\}$ or $k_{S_2}(b) \in \{3, 4\}$. □

5.3.5 Measures of Perturbation Versus Selected Proximity Measures of Multisets

Again, let us consider two multisets S_1 and S_2, such that $S_1, S_2 \in V^m(L)$, drawn from the set $V = \{v_1, v_2, \ldots, v_L\}$. The comparison of proximity measures of two multisets can be performed:

- *analytically* - where two measures are considered equivalent or one measure is expressed as a function of the other measure, or
- *empirically* - for a given data set.

Both cases are discussed below.

The measures known in the literature can be expressed as some functions of the measures of perturbation type 1 of one multiset by another multiset (cf. Krawczak and Szkatuła [10, 11]), or the measures of perturbation type 2 (cf. Krawczak and Szkatuła [12]). These measures can be split separated into two components, which correspond to the pair of the directional perturbations. In the following corollaries we present several very important properties of the selected measures, in which our idea of the multiset perturbation measures is involved.

At the beginning, we consider the so-called Bray-Curtis dissimilarity, Eq. 5.3, which can be explained in such a way that the equivalent definition contains the sum of the directional measures of the perturbation type 1.

Corollary 5.12 *The sum of the measures of perturbation type 1 satisfies the following condition*

$$D_{Bray-Curtis}(S_1, S_2) = Per^1_{MS}(S_1 \mapsto S_2) + Per^1_{MS}(S_2 \mapsto S_1)$$

Proof By Definition 5.2 and Eq. 5.3.

□

Next, likewise, the equivalent definition of the Steinhaus distance, Eq. 5.4, can be obtained as follows.

Corollary 5.13 *The sum of the measures of perturbation type 2 satisfies the following condition*

$$D_{Steinhaus}(S_1, S_2) = Per^2_{MS}(S_1 \mapsto S_2) + Per^2_{MS}(S_2 \mapsto S_1).$$

Proof By Definition 5.3 and Eq. 5.4.

□

Thus, the introduced measures of perturbations of one multiset by another multiset can be used to provide

Example 5.13 Let us assume that $L = 2$, and let us consider two exemplary multisets

$$S_1 = \{(k_{S_1}(a), a), (k_{S_1}(b), b)\},$$

$$S_2 = \{(k_{S_2}(a), a), (k_{S_2}(b), b)\},$$

drawn from the set $V = \{a, b\}$, where $S_1, S_2 \in V^5(2)$.

It is assumed additionally, that

$$k_{S_1}(a) = 2, \quad k_{S_1}(b) = 3 \text{ and}$$

$$k_{S_2}(a) = 3, \quad k_{S_2}(b) = 1.$$

So, we consider the pair of multisets

$$S_1 = \{(2, a), (3, b)\} \text{ and } S_2 = \{(3, a), (1, b)\}.$$

The multisets S_1 and S_2 can be represented as points in two-dimensional space, specified by the coordinates $k(a)$ and $k(b)$, namely as points (2,3) and (3,1), respectively. And then, we face the problem of calculation of the degrees of proximity between these two multisets.

Let us compare the proposed perturbations of one multiset S_1 by another multiset S_2 (and vice versa) to three commonly used distance measures, namely

$$D_{Chebyshev}(S_1, S_2), \text{ see Eq. 5.5,}$$

$$D_{Manhattan}(S_1, S_2), \text{ see Eq. 5.6,}$$

$$D_{Euclidean}(S_1, S_2), \text{ see Eq. 5.7.}$$

According to (5.8) and (5.9), the perturbation of the multiset S_2 by the multiset S_1 has the form

$$(S_1 \mapsto S_2) = \{(max\{k_{S_1}(a) - k_{S_2}(a), 0\}, a), (max\{k_{S_1}(b) - k_{S_2}(b), 0\}, b)\} =$$

$$= \{(0, a), (k_{S_1 \mapsto S_2}(b), b)\} = \{(0, a), (2, b)\},$$

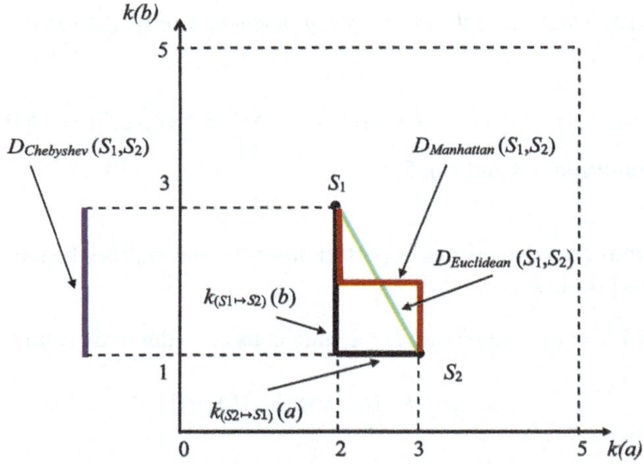

Fig. 5.4 The graphical illustration of the selected measures for fixed multisets S_1 and S_2

and the perturbation of the multiset S_1 by the multiset S_2 has the form

$$(S_2 \mapsto S_1) = \{(max\{k_{S_2}(a) - k_{S_1}(a), 0\}, a), (max\{k_{S_2}(b) - k_{S_1}(b), 0\}, b)\} =$$

$$= \{(k_{S_2 \mapsto S_1}(a), a), (0, b)\} = \{(1, a), (0, b)\}.$$

The values of the nonzero counting functions of the perturbations are

$$k_{S_1 \mapsto S_2}(b) = 2 \text{ and } k_{S_2 \mapsto S_1}(a) = 1.$$

The graphical illustration of the selected measures and the counting functions of proposed perturbations, for the fixed multisets S_1 and S_2, is shown in Fig. 5.4.

It is clear that the different criteria of evaluation of distances between multisets will lead to different results. Obviously, the Chebyshev measure

$$D_{Chebyshev}(S_1, S_2) = 2 \text{ (the purple segment)},$$

as well as the Manhattan distance

$$D_{Manhattan}(S_1, S_2) = 3 \text{ (the red path shows one of possible realizations)},$$

and the Euclidean distance

$$D_{Euclidean}(S_1, S_2) = \sqrt{5} \text{ (green segment)},$$

are symmetric. However, if the direction of comparison of multisets cannot be neglected, then the counting functions of the perturbations

$$k_{S_1 \mapsto S_2}(b) = 2 \text{ and } k_{S_2 \mapsto S_1}(a) = 1 \text{ (two black segments)}$$

may be used.

□

Thus, it is difficult to unequivocally decide which measure is better in general. In other words, there does not exist the best measure for evaluation of proximity between two arbitrary multisets and the choice depends on the nature of data under consideration.

5.4 Decomposition of Matrix of Measures

Often, in experimental studies, asymmetric pairwise relationships are observed which may be presented in the form of an asymmetric matrix of the applied different measures. However, asymmetry can be viewed as a combination of a symmetric component and an asymmetric component. If the asymmetry in the data is assumed to constitute noise, the asymmetric component can be ignored and the symmetric component can be analyzed alone.

Thus, a square asymmetric matrix

$$D = [d_{ij}],$$

for $i, j = 1, 2, ..., N$ (i.e., with N rows and N columns) can be additively decomposed into *a symmetric matrix* SYM and *a skew-symmetric matrix* $ASYM$ [29]. So, the relationship

$$D = SYM + ASYM$$

is fulfilled, where $SYM = [sym_{ij}]$ is a symmetric matrix of averages

$$sym_{ij} = \frac{d_{ij} + d_{ji}}{2}, \tag{5.14}$$

and $ASYM = [asym_{ij}]$ is a skew-symmetric matrix with elements

$$asym_{ij} = \frac{d_{ij} - d_{ji}}{2}, \tag{5.15}$$

for $i, j = 1, 2, ..., N$.

5.4.1 Decomposition of the Matrix of Multiset Perturbation Measures

Let us consider the multisets $S_i, S_j \in V^m(L)$, where $V = \{v_1, v_2, \dots, v_L\}$, for $i, j = 1, 2, ...N$, and m specifies the number of occurrences of the element $v \in V$ in the multiset. The matrix of some measures of multiset perturbations D_{Per} is shown in Table 5.1.

The elements of the matrix D_{Per}, according to Definition 5.2, have the following form:

$$d_{ij} = Per^1_{MS}(S_i \mapsto S_j) = \frac{\sum\limits_{l=1}^{L}(k_{S_i}(v_l) - k_{S_i \cap S_j}(v_l))}{\sum\limits_{l=1}^{L}(k_{S_i}(v_l) + k_{S_j}(v_l))}, \tag{5.16}$$

$$d_{ji} = Per^1_{MS}(S_j \mapsto S_i) = \frac{\sum\limits_{l=1}^{L}(k_{S_j}(v_l) - k_{S_j \cap S_i}(v_l))}{\sum\limits_{l=1}^{L}(k_{S_j}(v_l) + k_{S_i}(v_l))}, \tag{5.17}$$

for $i, j = 1, 2, ..., N$. Note that this matrix does not have to be symmetric, that is, the corresponding elements d_{ij} and d_{ji} do not have to be equal.

The matrix D_{Per} can be additively decomposed into two matrices, namely

$$D_{Per} = SYM_{Per} + ASYM_{Per}.$$

According to Eqs. 5.14 and 5.15, we obtain the following dependence

$$d_{ij} = sym_{ij} + asym_{ij},$$

Table 5.1 The matrix $D_{Per} = [d_{ij}]$, $i, j = 1, 2, ...N$

	S_1	S_2	...	S_i	...	S_j	...	S_N
S_1	0	d_{12}	...	d_{1i}	...	d_{1j}	...	d_{1N}
S_2	d_{21}	0	...	d_{2i}	...	d_{2j}	...	d_{2N}
$\vdots$	$\vdots$	$\vdots$	$\ddots$	$\vdots$	$\ddots$	$\vdots$	$\ddots$	$\vdots$
S_i	d_{i1}	d_{i2}	...	0	...	$\boldsymbol{d}_{ij}$	...	d_{iN}
$\vdots$	$\vdots$	$\vdots$	$\ddots$	$\vdots$	$\ddots$	$\vdots$	$\ddots$	$\vdots$
S_j	d_{j1}	d_{j2}	...	$\boldsymbol{d}_{ji}$	...	0	...	d_{jN}
$\vdots$	$\vdots$	$\vdots$	$\ddots$	$\vdots$	$\ddots$	$\vdots$	$\ddots$	$\vdots$
S_N	d_{N1}	d_{N2}	...	d_{Ni}	...	d_{Nj}	...	0

where

$$sym_{ij} = \frac{\sum_{l=1}^{L}(k_{S_i}(v_l) - k_{S_i \cap S_j}(v_l)) + \sum_{l=1}^{L}(k_{S_j}(v_l) - k_{S_j \cap S_i}(v_l))}{2\sum_{l=1}^{L}(k_{S_i}(v_l) + k_{S_j}(v_l))} =$$

$$= \frac{\sum_{l=1}^{L}(k_{S_i}(v_l) + k_{S_j}(v_l) - 2k_{S_i \cap S_j}(v_l))}{2\sum_{l=1}^{L}(k_{S_i}(v_l) + k_{S_j}(v_l))} \tag{5.18}$$

and

$$asym_{ij} = \frac{\sum_{l=1}^{L}(k_{S_i}(v_l) - k_{S_i \cap S_j}(v_l)) - \sum_{l=1}^{L}(k_{S_j}(v_l) - k_{S_j \cap S_i}(v_l))}{2\sum_{l=1}^{L}(k_{S_i}(v_l) + k_{S_j}(v_l))} =$$

$$= \frac{\sum_{l=1}^{L}(k_{S_i}(v_l) - k_{S_j}(v_l))}{2\sum_{l=1}^{L}(k_{S_i}(v_l) + k_{S_j}(v_l))}. \tag{5.19}$$

It is easy to see that the following conditions

$$0 \le sym_{ij} \le \frac{1}{2},$$

$$-\frac{1}{2} \le asym_{ij} \le \frac{1}{2},$$

are satisfied.

Let us consider two distinguished values (elements) of matrix D_{Per}, corresponding to the measure of perturbation $(S_i \mapsto S_j)$ (i.e., d_{ij}), and corresponding to the measure of perturbation $(S_j \mapsto S_i)$ (i.e., d_{ji}). Each of them can be decomposed into values:

$$sym_{ij}, asym_{ij} \quad \text{and} \quad sym_{ji}, asym_{ji},$$

respectively. Interesting relationships between them are presented below in the following corollaries.

Corollary 5.14 *The sum of the corresponding components, for the matrix $ASYM_{Per}$ satisfies the following condition*

$$asym_{ij} + asym_{ji} = 0.$$

Proof It should be noticed that according to Eq. 5.19 the following equality can be written

$$asym_{ij} + asym_{ji} =$$

$$= \frac{\sum_{l=1}^{L}(k_{S_i}(v_l) - k_{S_j}(v_l))}{2\sum_{l=1}^{L}(k_{S_i}(v_l) + k_{S_j}(v_l))} + \frac{\sum_{l=1}^{L}(k_{S_j}(v_l) - k_{S_i}(v_l))}{2\sum_{l=1}^{L}(k_{S_j}(v_l) + k_{S_i}(v_l))} = 0.$$

□

Therefore, the matrix $ASYM_{Per}$ is a skew-symmetric matrix. A different relationship can be formulated for the matrix SYM_{Per}.

Corollary 5.15 *The sum of the corresponding components for the matrix SYM_{Per} satisfies the following condition*

$$sym_{ij} + sym_{ji} = 1 - S_{Tversky}\left(S_i, S_j; \frac{1}{2}, \frac{1}{2}\right),$$

where $S_{Tversky}(S_i, S_j; \frac{1}{2}, \frac{1}{2})$ is Tversky's similarity measure for parameters $\alpha = \beta = \frac{1}{2}$.

Proof It should be noticed that according to Eq. 5.18 the following equality can be written

$$sym_{ij} + sym_{ji} =$$

$$= \frac{1}{2} - \frac{\sum_{l=1}^{L}(k_{S_i \cap S_j}(v_l))}{\sum_{l=1}^{L}(k_{S_i}(v_l) + k_{S_j}(v_l))} + \frac{1}{2} - \frac{\sum_{l=1}^{L}(k_{S_j \cap S_i}(v_l))}{\sum_{l=1}^{L}(k_{S_j}(v_l) + k_{S_i}(v_l))} =$$

$$= 1 - \frac{2\sum_{l=1}^{L}(k_{S_i \cap S_j}(v_l))}{\sum_{l=1}^{L}(k_{S_i}(v_l) + k_{S_j}(v_l))} = 1 - S_{Tversky}(S_i, S_j; \frac{1}{2}, \frac{1}{2}).$$

□

Thus, the sum of two related components of the matrix SYM_{Per}, one corresponding to the perturbation ($S_i \mapsto S_j$) (i.e., sym_{ij}), and the other, related to the perturbation ($S_j \mapsto S_i$) (i.e., sym_{ji}), gives an equivalent interpretation of dissimilarity of two multisets S_i and S_j.

5.4.2 Graphical Illustration of Decomposition of the Matrix of Multiset Perturbation Measures

In order to demonstrate the meaning of decomposition of the matrix of multiset perturbation measures we draw some possible graphical illustrations in two-dimensional space, by analyzing the following examples.

Example 5.14 Let us assume that $V = \{a\}$, that is, $L = card(V) = 1$, and consider the following multisets $S_i, S_j \in V^5(1)$, denoted by

$$S_i = \{(k_{S_i}(a), a)\} \text{ and } S_j = \{(k_{S_j}(a), a)\},$$

with $m = 5$. According to Eqs. 5.16 and 5.17, the measures of perturbation in the matrix D_{Per} have the following forms:

$$d_{ij} = Per^1_{MS}(S_i \mapsto S_j) = \frac{k_{S_i}(a) - k_{S_i \cap S_j}(a)}{k_{S_i}(a) + k_{S_j}(a)},$$

$$d_{ji} = Per^1_{MS}(S_j \mapsto S_i) = \frac{k_{S_j}(a) - k_{S_j \cap S_i}(a)}{k_{S_2}(a) + k_{S_1}(a)}.$$

According to Eqs. 5.18 and 5.19, the element d_{ij} can be decomposed into two elements

$$sym_{ij} = \frac{k_{S_i}(a) + k_{S_j}(a) - 2k_{S_i \cap S_2}(a)}{2(k_{S_i}(a) + k_{S_j}(a))},$$

$$asym_{ij} = \frac{k_{S_i}(a) - k_{S_j}(a)}{2(k_{S_i}(a) + k_{S_j}(a))},$$

and the element d_{ji} can be decomposed into

$$sym_{ji} = \frac{k_{S_j}(a) + k_{S_i}(a) - 2k_{S_j \cap S_i}(a)}{2(k_{S_j}(a) + k_{S_i}(a))},$$

$$asym_{ji} = \frac{k_{S_j}(a) - k_{S_i}(a)}{2(k_{S_j}(a) + k_{S_i}(a))}.$$

Additionally, it is assumed, that the counting function for the multiset S_i is being changed from 0 to 5, that is

$$k_{S_i}(a) \in \{0, 1, ..., 5\},$$

while the counting function for the multiset S_j is also being changed from 0 to 5, that is

$$k_{S_j}(a) \in \{0, 1, ..., 5\}.$$

In this way, we can consider the ordered pairs of the multisets (S_i, S_j), for $i, j \in \{0, 1, ..., 5\}$ as follows

$$(S_0, S_0),\ (S_0, S_1),\ (S_0, S_2),\ ...,\ (S_0, S_5)$$

$$(S_1, S_0),\ (S_1, S_1),\ (S_1, S_2),\ ...,\ (S_1, S_5)$$

$$...$$

$$(S_5, S_0),\ (S_5, S_1),\ (S_5, S_2),\ ...,\ (S_5, S_5)$$

where

$$S_0 = \{(0, a)\},\ S_1 = \{(1, a)\},\ S_2 = \{(2, a)\},$$

$$S_3 = \{(3, a)\},\ S_4 = \{(4, a)\},\ S_5 = \{(5, a)\}.$$

According to Eq. 5.16 the measures of perturbations $Per^1_{MS}(S_i \mapsto S_j)$, for $i, j \in \{0, 1, ..., 5\}$ in the matrix D_{Per} take the values as shown in Table 5.2.

Figure 5.5 shows the values of the measures of the perturbation

$$d_{ij} = Per^1_{MS}(S_i \mapsto S_j) \text{ (the abscissa axis)},$$

for the different ordered pairs

$$(S_i, S_j) \text{ for } i, j = 0, 1, 2, ..., 5,$$

where the value of $k_{S_i}(a)$ indicates the curve in order, and $k_{S_j}(a)$ is changing from 0 to 5 (the ordinate axis). The values of the measures are indicated as bold points on the lines.

Table 5.2 Example of the matrix D_{Per}

	S_0	S_1	S_2	S_3	S_4	S_5
S_0	0	0	0	0	0	0
S_1	1	0	0	0	0	0
S_2	1	0.33	0	0	0	0
S_3	1	0.50	0.20	0	0	0
S_4	1	0.60	0.33	0.14	0	0
S_5	1	0.66	0.42	0.25	0.11	0

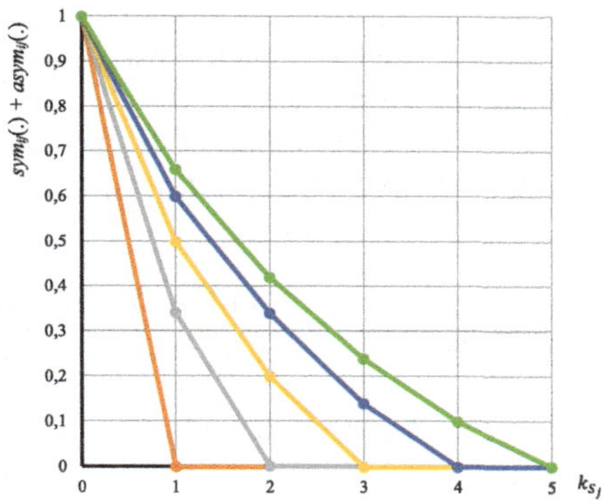

Fig. 5.5 The measures of perturbations, where each value of $k_{S_i}(a)$ indicates the curve in order, $i = 0, 1, 2, ..., 5$, and the value of $k_{S_j}(a)$ is changed from 0 to 5 (the ordinate axis)

Table 5.3 The matrix SYM_{Per}

	S_0	S_1	S_2	S_3	S_4	S_5
S_0	0	0.5	0.5	0.5	0.5	0.5
S_1	0.5	0	0.17	0.25	0.30	0.33
S_2	0.5	0.17	0	0.10	0.17	0.21
S_3	0.5	0.25	0.10	0	0.07	0.12
S_4	0.5	0.30	0.17	0.07	0	0.05
S_5	0.5	0.33	0.21	0.12	0.05	0

Table 5.4 The matrix $ASYM_{Per}$

	S_0	S_1	S_2	S_3	S_4	S_5
S_0	0	−0.5	−0.5	−0.5	−0.5	−0.5
S_1	0.5	0	−0.17	−0.25	−0.30	−0.33
S_2	0.5	0.17	0	−0.10	−0.17	−0.21
S_3	0.5	0.25	0.10	0	−0.07	−0.12
S_4	0.5	0.30	0.17	0.07	0	−0.05
S_5	0.5	0.33	0.21	0.12	0.05	0

The matrix D_{Per} can be decomposed as a sum of two matrices:

- a symmetric matrix SYM_{Per}, and
- a skew-symmetric matrix $ASYM_{Per}$,

as shown in Tables 5.3 and 5.4. It is interesting to note that the asymmetric part $ASYM_{Per}$ takes some negative values.

It is shown in Fig. 5.6 that the measures of perturbation d_{ij} can be divided into symmetric part sym_{ij} (left figure) and asymmetric part $asym_{ij}$ (right figure), where the value of $k_{S_i}(a)$ indicates the curve in order, $i = 0, 1, 2, ..., 5$, and $k_{S_j}(a)$ is changed from 0 to 5 (the ordinate axis). The exact values of the measures are indicated as the bold points on the lines. The condition $sym_{ij} + asym_{ij} = d_{ij}$ is satisfied.

□

For example, we can trace one curve from Fig. 5.5 and the two corresponding curves in Fig. 5.6. Assuming that

$$k_{S_i}(a) = 2, \text{ and}$$

$$k_{S_j}(a) \text{ changes from 0 to 5 (the ordinate axis),}$$

the respective result is shown in Figs. 5.7 and 5.8. The bold points on the lines indicate the values of the measures.

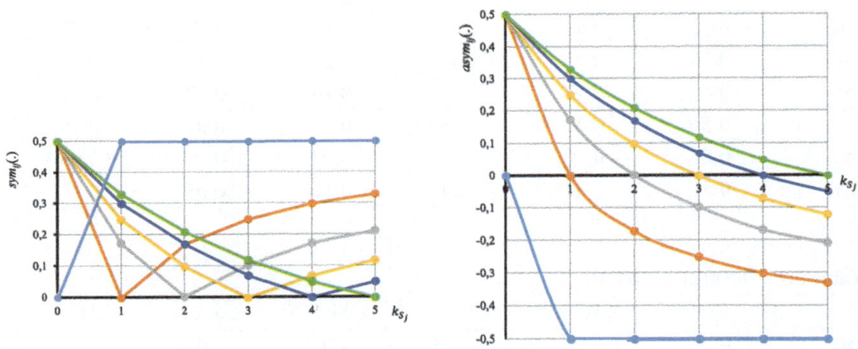

Fig. 5.6 The values of the measures of perturbations divided into symmetric part (left figure) and asymmetric part (right figure), where the value of $k_{S_i}(a)$ indicates the curve in order, $i = 0, 1, 2, ..., 5$, and $k_{S_j}(a)$ is changed from 0 to 5 (the ordinate axis)

Fig. 5.7 The measures of perturbations, where the value of $k_{S_i}(a) = 2$, and the value of $k_{S_j}(a)$ is changed from 0 to 5 (the ordinate axis)

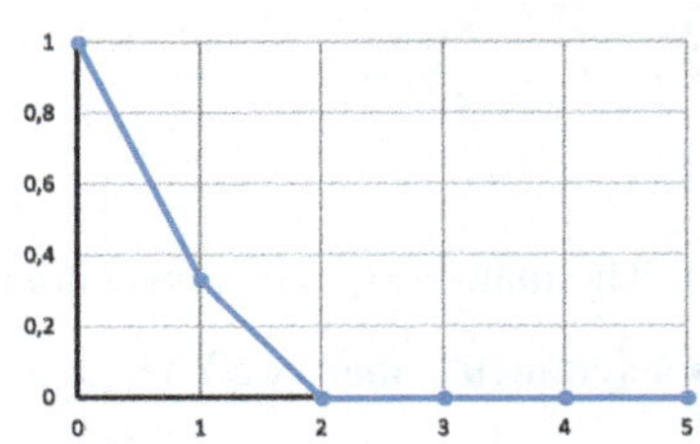

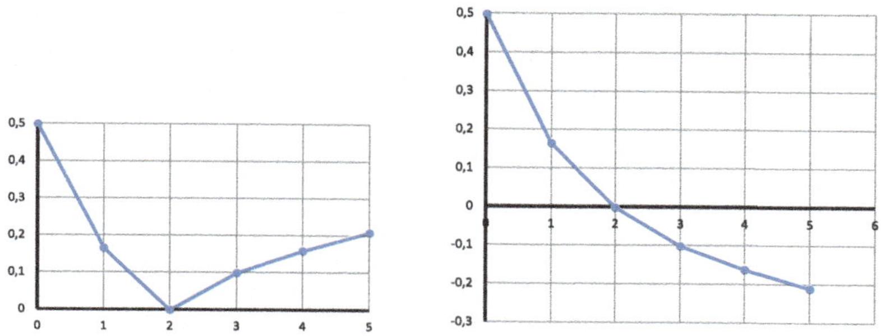

Fig. 5.8 The values of the measures of perturbations divided into symmetric part (left figure) and asymmetric part (right figure), where the value of $k_{S_i}(a) = 2$, and $k_{S_j}(a)$ is changed from 1 to 5 (the ordinate axis)

5.5 Representation of Multi-attribute Objects by Multisets

Until now we have been dealing, in this chapter, with the fundamental definitions, concerning the perturbation of multisets (cf. Krawczak and Szkatuła [26]). Now, we will describe a multi-attribute object with repeating nominal values of attributes, understood as an ordered collection of multisets. Next, the concept of the measure of perturbation of one multiset by another multiset is adapted to other multisets, which describe the object and the group of such objects.

Let us consider a collection of some multi-attribute qualitative objects $U = \{e_n\}$, indexed by n, for $n = 1, 2, ..., N$. The objects are described by K attributes $A = \{a_1, a_2, ..., a_K\}$, indexed by j, for $j = 1, 2, ..., K$. The set

$$V_{a_j} = \{v_{1,j}, v_{2,j}, ..., v_{L_j,j}\}$$

is the domain of the attribute $a_j \in A$, $j = 1, 2, ..., K$, where L_j denotes the number of nominal values of the attribute a_j.

For the above we assume that the considered multi-attribute objects can be characterized by repeated values of the attributes. Additionally, we have extra information as to, how many times each value $v_{i,j} \in V_{a_j}$, for $i = 1, 2, ..., L_j$ and $j = 1, 2, ..., K$, is repeated for each object $e_n \in U$, where the number j determines the considered attribute a_j.

5.5.1 *Description of Multi-attribute Objects by Multisets*

We assume that each object is represented by its description. *The description of an object e* is denoted by G_e, and can be represented by an ordered collection of multisets, see the following definition.

Definition 5.4 (*Description of an object*) Every object $e \in U$ can be represented by an ordered collection of K multisets $S_{j,t(j,e)}$, $j = 1, 2, ..., K$, drawn from the ordinary sets of nominal values $V_{a_j} = \{v_{1,j}, v_{2,j}, ..., v_{L_j,j}\}$ of the attributes a_j, given as follows

$$G_e =< S_{1,t(1,e)},\ S_{2,t(2,e)}, ..., S_{K,t(K,e)} > \tag{5.20}$$

where

$$1 \leq card(S_{j,t(j,e)}) \leq m,$$

for

$$S_{j,t(j,e)} \in V_{a_j}^m(L_j),$$

$j \in \{1, 2, ..., K\}$, and m is an integer number indicated in Definition 5.1.

In Definition 5.4, the description of an object e is denoted by G_e, while each multiset $S_{j,t(j,e)}$, constituting it, represents the attribute a_j, where $j = 1, 2, ..., K$. Hence, the subscript

$$j, t(j, e), \text{ for } j = 1, 2, ..., K,$$

specifies the attribute a_j of the object e, while the multiset $S_{j,t(j,e)}$ represents the description of this attribute. Each j-th multiset $S_{j,t(j,e)}$ (the number j specifies that attribute a_j is considered) can be represented by the set of L_j pairs,

$$S_{j,t(j,e)} = \{(k_{S_{j,t(j,e)}}(v_{i,t(j,e)}),\ v_{i,t(j,e)}) \mid i = 1, 2, ..., L_j\} =$$

$$= \{(k_{S_{j,t(j,e)}}(v_{1,t(j,e)}),\ v_{1,t(j,e)}), ..., k_{S_{j,t(j,e)}}(v_{L_j,t(j,e)}),\ v_{L_j,t(j,e)})\}, \tag{5.21}$$

where $v_{i,t(j,e)} \in V_{a_j}$, for $j = 1, 2, ..., K$. The value

$$k_{S_{j,t(j,e)}}(v_{i,t(j,e)}), \text{ for } i = 1, 2, ..., L_j,$$

specifies the number of occurrences of the value $v_{i,t(j,e)} \in V_{a_j}$ in the multiset $S_{j,t(j,e)}$. Another subscript

$$i, t(j, e)$$

specifies which element $v_{i,t(j,e)}$ from the set

$$V_{a_j} = \{v_{1,j}, v_{2,j}, ..., v_{L_J,j}\}$$

for the attribute a_j, and for object e, is considered.

Thus, the applied notation states that for the object e, and for the attribute a_j, the value

$$v_{1,t(j,e)} \in V_{a_j} \text{ appears } k_{S_{j,t(j,e)}}(v_{1,t(j,e)}) \text{ times},$$

and the value

$$v_{2,t(j,e)} \in V_{a_j} \text{ appears } k_{S_{j,t(j,e)}}(v_{2,t(j,e)}) \text{ times,}$$

and so on. Thus, it is obvious that each multiset $S_{j,t(j,e)}$ represents the separate attribute a_j, which takes the values $v_{i,t(j,e)} \in V_{a_j}$, $j = 1, 2, ..., K$.

Let us analyse the following example that will explain the intricacies of the introduced notation.

Example 5.15 Consider the object e described by two attributes $\{a_1, a_2\}$, where the sets

$$V_{a_1} = \{v_{1,1},\ v_{2,1},\ v_{3,1}\} \text{ and } V_{a_2} = \{v_{1,2},\ v_{2,2}\}$$

are the domains of these attributes, respectively. According to (5.20), the object e can be described by an ordered collection of two multisets

$$G_e =< S_{1,t(1,e)},\ S_{2,t(2,e)} > .$$

Let us assume that due to (5.21), the exemplary multisets $S_{1,t(1,e)}$ and $S_{2,t(2,e)}$ have the form

$$S_{1,t(1,e)} = \{(2,\ v_{1,1}),\ (1,\ v_{3,1})\} \text{ and } S_{2,t(2,e)} = \{(2,\ v_{1,2})\}.$$

Thus, the description of the object e can be written in the following multiset form

$$G_e =< \{(2,\ v_{1,1}), (1,\ v_{3,1})\},\ \{(2,\ v_{1,2})\} > .$$

□

The arithmetic addition of the multisets (denoted by $\oplus$) yields a new multiset, and can be applied to all multisets in the descriptions G_{e_1} and G_{e_2}. Let us consider two objects e_1 and e_2, and their descriptions

$$G_{e_1} =< S_{1,t(1,e_1)},\ S_{2,t(2,e_1)}, ...,\ S_{K,t(K,e_1)} >,$$

$$G_{e_2} =< S_{1,t(1,e_2)},\ S_{2,t(2,e_2)}, ...,\ S_{K,t(K,e_2)} > .$$

We can now introduce the definition of the join between the descriptions of objects.

Definition 5.5 (*Join between descriptions of objects*) The join between the description of an object e_1 and the description of an object e_2 has the following description

$$G_{e_1} \oplus G_{e_2} =$$

$$=< S_{1,t(1,e_1)} \oplus S_{1,t(1,e_2)},\ S_{2,t(2,e_1)} \oplus S_{2,t(2,e_2)}, ...,\ S_{K,t(K,e_1)} \oplus S_{K,t(K,e_2)} > . \quad (5.22)$$

The definition says that the description of two joined objects is represented by a new collection of K multisets. Each such j-th multiset, $j = \{1, 2, ..., K\}$, is constructed as the join of two multisets

$$S_{j,t(j,e_1)} \oplus S_{j,t(j,e_2)}$$

describing the attribute a_j for the objects e_1 and e_2, respectively.

A short consideration of a special case will explain the intricacies of the above definition.

Exemplary case for $K = 1$

Now, let us consider a special case for $K = 1$ (i.e., an object e is described by a single attribute $A = \{a_1\}$), where the set

$$V_{a_1} = \{v_{1,1}, v_{2,1}, ..., v_{L_1,1}\}$$

is the domain of this attribute. Each object e can be represented by a single multiset $S_{1,t(1,e)}$ drawn from the ordinary set of values V_{a_1}. In this case, the description of each object e, defined in (5.20), is reduced to the form

$$G_e =< S_{1,t(1,e)} >,$$

where $S_{1,t(1,e)} \in V_{a_1}^m(L_1)$ and is defined by (5.21), and now can be rewritten in the following form

$$S_{1,t(1,e)} = \{(k_{S_{1,t(1,e)}}(v_{1,t(1,e)}), v_{1,t(1,e)}), ..., k_{S_{1,t(1,e)}}(v_{L_1,t(1,e)}), v_{L_1,t(1,e)})\}, \quad (5.23)$$

where $v_{i(1),t(1,e)} \in V_{a_1}$, for $i(1) \in \{1, 2, ..., L_1\}$.

The index

$$i(1), t(1, e)$$

specifies which value $v_{i(1),t(1,e)}$ of the attribute a_1 is used in the object e. For the object e and for the attribute a_1 the value $v_{i(1),t(1,e)}$ appears $k_{S_{1,t(1,e)}}(v_{i(1),t(1,e)})$ times in the multiset $S_{1,t(1,e)}$, for $i(1) \in \{1, 2, ..., L_1\}$.

Thus, it is a proper time for presenting the details of the proposed approach concerning the measure of perturbation of one object by another object.

5.5.2 Measure of Perturbation of Objects

Consider two objects $e_1, e_2 \in U$, described by K attributes $A = \{a_1, a_2, ..., a_K\}$, where the set $V_{a_j} = \{v_{1,j}, v_{2,j}, ..., v_{L_j,j}\}$ is the domain of the attribute $a_j \in A$, $j = 1, 2, ..., K$. According to (5.20), the respective descriptions of the objects are as follows:

$$G_{e_1} =< S_{1,t(1,e_1)}, S_{2,t(2,e_1)}, ..., S_{K,t(K,e_1)} >,$$

$$G_{e_2} =< S_{1,t(1,e_2)}, S_{2,t(2,e_2)}, ..., S_{K,t(K,e_2)} >,$$

where $S_{j,t(j,e_1)}, S_{j,t(j,e_2)} \in V^m(L_j)$, for $j = 1, 2, ..., K$.

The new concept of object perturbation is defined as follows.

Definition 5.6 (*Perturbation of object*) The perturbation of the object e_2 by the object e_1, denoted by $(G_{e_1} \mapsto G_{e_2})$, can be represented by an ordered collection of multisets $S_{j,t(j,e_1)} \ominus S_{j,t(j,e_2)}$, $j = 1, 2, ..., K$, drawn from the ordinary sets of nominal values V_{a_j} of the attributes a_j, respectively

$$(G_{e_1} \mapsto G_{e_2}) =$$

$$= < (S_{1,t(1,e_1)} \mapsto S_{1,t(1,e_2)}), \; ..., \; (S_{K,t(K,e_1)} \mapsto S_{K,t(K,e_2)}) > =$$

$$= < S_{1,t(1,e_1)} \ominus S_{1,t(1,e_2)}, \; S_{2,t(2,e_1)} \ominus S_{2,t(2,e_2)}, \; ..., \; S_{K,t(K,e_1)} \ominus S_{K,t(K,e_2)} > \tag{5.24}$$

Thus, the perturbation of the object e_2 by the object e_1 is represented by the collection of the multisets, where each such multiset is constructed as the difference of two multisets (denoted by $\ominus$), describing the attribute a_j, $j = 1, 2, ..., K$, of the object e_2 and the object e_1.

The counterpart case is defined in an analogous manner, that is

$$(G_{e_2} \mapsto G_{e_1}) =$$

$$= < (S_{1,t(1,e_2)} \mapsto S_{1,t(1,e_1)}), ..., (S_{K,t(K,e_2)} \mapsto S_{K,t(K,e_1)}) > =$$

$$= < (S_{1,t(1,e_2)} \ominus S_{1,t(1,e_1)}), \; ..., \; (S_{K,t(K,e_2)} \ominus S_{K,t(K,e_1)}) > \tag{5.25}$$

So, the measure of perturbation of one object by another object is a number ranging between 0 and 1 and obtained via some aggregation operation. The aggregation is done on a set of measures of the perturbations associated with each attribute a_j, for $j = 1, 2, ..., K$, see the following Definition 5.7.

Definition 5.7 (*Measure of perturbation of object*) The measure of perturbation of the object e_2 by the object e_1, denoted by $Per_O(G_{e_1} \mapsto G_{e_2})$, is defined in the following manner:

$$Per_O(G_{e_1} \mapsto G_{e_2}) =$$

$$= Agg(Per_{MS}(S_{1,t(1,e_1)} \mapsto S_{1,t(1,e_2)}), ..., Per_{MS}(S_{K,t(K,e_1)} \mapsto S_{K,t(K,e_2)})) \tag{5.26}$$

where Agg is an aggregation operation.

In the opposite case, the measure of the perturbation of the object e_1 by the object e_2 is defined analogously:

$$Per_O(G_{e_2} \mapsto G_{e_1}) =$$

$$= Agg(Per_{MS}(S_{1,t(1,e_2)} \mapsto S_{1,t(1,e_1)}), ..., Per_{MS}(S_{K,t(K,e_2)} \mapsto S_{K,t(K,e_1)})) \tag{5.27}$$

The aggregation operation used in (5.26) and (5.27) is described as a mapping $Agg : [0, 1]^K \rightarrow [0, 1]$, which assigns a real number to any K-tuple $(p_1, p_2, ..., p_K)$ of real numbers.

The aggregation operation was discussed in more detail in Sect. 3.4.2. For example, the aggregation operation Agg can be realized by the function of weighted average of its arguments, that is

$$Agg(p_1, p_2, ..., p_K) = \frac{1}{K}\sum_{j=1}^{K}(w_j \cdot p_j),$$

where the weight parameter $w_j > 0$ determines the importance of the jth element, for $j = 1, 2, ..., K$.

According to (5.26), the measure of perturbation of the object e_2 by the object e_1 may be rewritten in the following manner for the measure of perturbation type 1 (the index 1, indicating the type of perturbation was omitted)

$$Per_O(G_{e_1} \mapsto G_{e_2}) = \frac{1}{K}\sum_{j=1}^{K}(w_j \cdot Per_{MS}(S_{j,t(j,e_1)} \mapsto S_{j,t(j,e_2)})) =$$

$$= \frac{1}{K}\sum_{j=1}^{K}(w_j \cdot \frac{\sum_{i=1}^{L_j}(k_{S_{j,t(j,e_1)}}(v_i) - k_{S_{j,t(j,e_1)} \cap S_{j,t(j,e_2)}}(v_i))}{\sum_{i=1}^{L_j}(k_{S_{j,t(j,e_1)}}(v_i) + k_{S_{j,t(j,e_2)}}(v_i))}). \tag{5.28}$$

In the opposite case, the measure of perturbation of the object e_1 by the object e_2 is rewritten as,

$$Per_O(G_{e_2} \mapsto G_{e_1}) = \frac{1}{K}\sum_{j=1}^{K}(w_j \cdot Per_{MS}(S_{j,t(j,e_2)} \mapsto S_{j,t(j,e_1)})) =$$

$$= \frac{1}{K}\sum_{j=1}^{K}(w_j \cdot \frac{\sum_{i=1}^{L_j}(k_{S_{j,t(j,e_2)}}(v_i) - k_{S_{j,t(j,e_2)} \cap S_{j,t(j,e_1)}}(v_i))}{\sum_{i=1}^{L_j}(k_{S_{j,t(j,e_2)}}(v_i) + k_{S_{j,t(j,e_1)}}(v_i))}). \tag{5.29}$$

For further considerations, let us assume that $w_j = 1$, for $j = 1, 2, ..., K$.

For these conditions, we can prove some properties of the measure of the object perturbations, which are included in the following corollaries: Corollaries 5.16, 5.17 and 5.18.

Corollary 5.16 *The measure of perturbation of the object e_2 by the object e_1, represented by respective descriptions G_{e_2} and G_{e_1}, satisfies the following inequality*

$$0 \leq Per_O(G_{e_1} \mapsto G_{e_2}) \leq 1. \tag{5.30}$$

Proof (1) First, we prove the left hand side inequality $Per_O(G_{e_1} \mapsto G_{e_2}) \geq 0$. It should be noticed that $\forall i \in \{1, 2, ..., L_j\}$, the inequality

$$k_{S_{j,t(j,e_1)}} \geq k_{S_{j,t(j,e_1)} \cap S_{j,t(j,e_2)}}(v_i),$$

for $j = 1, 2, ..., K$, is satisfied, and then

$$k_{S_{j,t(j,e_1)}} - k_{S_{j,t(j,e_1)} \cap S_{j,t(j,e_2)}}(v_i) \geq 0.$$

Due to Eq. 5.28 the following inequality can be written

$$Per_O(G_{e_1} \mapsto G_{e_2}) = \frac{1}{K} \sum_{j=1}^{K} \frac{\sum_{i=1}^{L_j} (k_{S_{j,t(j,e_1)}}(v_i) - k_{S_{j,t(j,e_1)} \cap S_{j,t(j,e_2)}}(v_i))}{\sum_{i=1}^{L_j} (k_{S_{j,t(j,e_1)}}(v_i) + k_{S_{j,t(j,e_2)}}(v_i))} \geq 0.$$

(2) Then, we prove the second inequality. It should be noticed that the inequality

$$k_{S_{j,t(j,e_1)}} - k_{S_{j,t(j,e_1)} \cap S_{j,t(j,e_2)}}(v_i) \leq k_{S_{j,t(j,e_1)}} + k_{S_{j,t(j,e_1)} \cap S_{j,t(j,e_2)}}(v_i),$$

$\forall i \in \{1, 2, ..., L_j\}$, for $j = 1, 2, ..., K$ is satisfied. Thus, the following inequality can be obtained

$$Per_O(G_{e_1} \mapsto G_{e_2}) = \frac{1}{K} \sum_{j=1}^{K} \frac{\sum_{i=1}^{L_j} (k_{S_{j,t(j,e_1)}}(v_i) - k_{S_{j,t(j,e_1)} \cap S_{j,t(j,e_2)}}(v_i))}{\sum_{i=1}^{L_j} (k_{S_{j,t(j,e_1)}}(v_i) + k_{S_{j,t(j,e_2)}}(v_i))} \leq$$

$$\leq \frac{1}{K} \sum_{j=1}^{K} \frac{\sum_{i=1}^{L_j} (k_{S_{j,t(j,e_1)}}(v_i) + k_{S_{j,t(j,e_2)}}(v_i))}{\sum_{i=1}^{L_j} (k_{S_{j,t(j,e_1)}}(v_i) + k_{S_{j,t(j,e_2)}}(v_i))} = 1.$$

□

Corollary 5.17 *The sum of the measures of perturbation $Per_O(G_{e_1} \mapsto G_{e_2})$ and $Per_O(G_{e_2} \mapsto G_{e_1})$ satisfies the following inequality*

$$0 \leq Per_O(G_{e_1} \mapsto G_{e_2}) + Per_O(G_{e_2} \mapsto G_{e_1}) \leq 1. \tag{5.31}$$

Proof (1) First, we prove the left hand side inequality. According to Eq. 5.30, (i.e., the inequalities $0 \leq Per_O(G_{e_1} \mapsto G_{e_2})$ and $0 \leq Per_O(G_{e_2} \mapsto G_{e_1})$ are satisfied), we obtain the following inequality

$$Per_O(G_{e_1} \mapsto G_{e_2}) + Per_O(G_{e_2} \mapsto G_{e_1}) \geq 0.$$

(2) The second inequality can be proved in the following way. One can notice that each inequality

$$k_{S_{j,t(j,e_1)} \cap S_{j,t(j,e_2)}}(v_i) \geq 0, \quad \forall i \in \{1, 2, ..., L_j\},$$

where $j = 1, 2, ..., K$, is satisfied. Due to Eqs. 5.28 and 5.29, one can obtain the right hand side inequality in the following way

$$Per_O(G_{e_1} \mapsto G_{e_2}) + Per_O(G_{e_2} \mapsto G_{e_1}) =$$

$$= \frac{1}{K} \sum_{j=1}^{K} \frac{\sum_{i=1}^{L_j} (k_{S_{j,t(j,e_1)}}(v_i) + k_{S_{j,t(j,e_2)}}(v_i) - 2 \cdot k_{S_{j,t(j,e_1)} \cap S_{j,t(j,e_2)}}(v_i))}{\sum_{i=1}^{L_j} (k_{S_{j,t(j,e_1)}}(v_i) + k_{S_{j,t(j,e_2)}}(v_i))} \leq$$

$$\leq \frac{1}{K} \sum_{j=1}^{K} \frac{\sum_{i=1}^{L_j} (k_{S_{j,t(j,e_1)}}(v_i) + k_{S_{j,t(j,e_2)}}(v_i))}{\sum_{i=1}^{L_j} (k_{S_{j,t(j,e_1)}}(v_i) + k_{S_{j,t(j,e_2)}}(v_i))} = 1.$$

□

Corollary 5.18 *The sum of the measures of perturbation $Per_O(G_{e_1} \mapsto G_{e_2})$ and $Per_O(G_{e_2} \mapsto G_{e_1})$ satisfies the following inequality*

$$Per_O(G_{e_1} \mapsto G_{e_2}) + Per_O(G_{e_2} \mapsto G_{e_1}) =$$

$$= 1 - \frac{1}{K} \sum_{j=1}^{K} \frac{2 \cdot \sum_{i=1}^{L_j} k_{S_{j,t(j,e_1)} \cap S_{j,t(j,e_2)}}(v_i)}{\sum_{i=1}^{L_j} (k_{S_{j,t(j,e_1)}}(v_i) + k_{S_{j,t(j,e_2)}}(v_i))}. \tag{5.32}$$

Proof Due to Definition 5.7 and Eqs. 5.28 and 5.29, the following equality can be obtained

$$Per_O(G_{e_1} \mapsto G_{e_2}) + Per_O(G_{e_2} \mapsto G_{e_1}) =$$

$$= \frac{1}{K}\sum_{j=1}^{K}\frac{\sum_{i=1}^{L_j}(k_{S_{j,t(j,e_1)}}(v_i) - k_{S_{j,t(j,e_1)}\cap S_{j,t(j,e_2)}}(v_i))}{\sum_{i=1}^{L_j}(k_{S_{j,t(j,e_1)}}(v_i) + k_{S_{j,t(j,e_2)}}(v_i))} +$$

$$+ \frac{1}{K}\sum_{j=1}^{K}\frac{\sum_{i=1}^{L_j}(k_{S_{j,t(j,e_2)}}(v_i) - k_{S_{j,t(j,e_2)}\cap S_{j,t(j,e_1)}}(v_i))}{\sum_{i=1}^{L_j}(k_{S_{j,t(j,e_2)}}(v_i) + k_{S_{j,t(j,e_1)}}(v_i))} =$$

$$= \frac{1}{K}\sum_{j=1}^{K}\frac{\sum_{i=1}^{L_j}(k_{S_{j,t(j,e_1)}}(v_i) + k_{S_{j,t(j,e_2)}}(v_i) - 2\cdot k_{S_{j,t(j,e_1)}\cap S_{j,t(j,e_2)}}(v_i))}{\sum_{i=1}^{L_j}(k_{S_{j,t(j,e_1)}}(v_i) + k_{S_{j,t(j,e_2)}}(v_i))} =$$

$$= 1 - \frac{1}{K}\sum_{j=1}^{K}\frac{2\cdot\sum_{i=1}^{L_j}k_{S_{j,t(j,e_1)}\cap S_{j,t(j,e_2)}}(v_i))}{\sum_{i=1}^{L_j}(k_{S_{j,t(j,e_1)}}(v_i) + k_{S_{j,t(j,e_2)}}(v_i))}.$$

□

Thus, the sum of the measure of perturbation of the object e_1 by the object e_2 and the measure of perturbation of the object e_2 by the object e_1 leads to a symmetric expression for the dissimilarity of two objects. In this way, Eq. 5.32 can be rewritten, and the equivalent definition of similarity of the objects can be obtained:

$$Similarity(G_{e_1}, G_{e_2}) = 1 - \left(Per_O(G_{e_1} \mapsto G_{e_2}) + Per_O(G_{e_2} \mapsto G_{e_1})\right),$$

which is based on our idea of the object perturbation measures.

5.5.3 Measure of Perturbation of Group of Objects

Let us assume that every non-empty subset of a finite set $U = \{e_n\}$, for $n = 1, 2, ..., N$, is called *a group*. We assume that *the description of the group* g is denoted by G_g. Let us consider a non-empty group of objects $g \subseteq U$ containing the objects $\{e_n : n \in J_g \subseteq \{1, 2, ...N\}\}$. According to (5.20), every object $e_n \in g$ can be represented by an ordered collection of multisets $S_{j,t(j,e_n)}$, for $j = 1, 2, ..., K$, drawn from the ordinary sets of values $V_{a_j} = \{v_{1,j}, v_{2,j}, ..., v_{L_j,j}\}$ of attributes a_j, that is

$$G_{e_n} =< S_{1,t(1,e_n)},\ S_{2,t(2,e_n)},\ ...,\ S_{K,t(K,e_n)} >,$$

for $S_{j,t(j,e_n)} \in V_{a_j}^m(L_j)$.

Thus, the group of objects g can be represented by an ordered collection of multisets, such that each multiset is drawn from the ordinary sets of values V_{a_j}, for $j = 1, 2, ..., K$, and the description of such a group is defined as follows,

$$G_g = \bigoplus_{n \in J_g} G_{e_n},$$

see Definition 5.8.

Definition 5.8 (*Description of group of objects*) A group of objects g, can be represented by an ordered collection of multisets $S_{j,t(j,g)}$, $j = 1, 2, ..., K$, drawn from the ordinary set of nominal values V_{a_j} of the attribute a_j, and is described as follows

$$G_g =< S_{1,t(1,g)}, S_{2,t(2,g)}, ..., S_{K,t(K,g)} > \quad (5.33)$$

where the multiset $S_{j,t(j,g)} \in V_{a_j}^{m_j}(L_j)$ for $j \in \{1, 2, ..., K\}$.

Hence, considering two groups of objects $g_1, g_2 \subseteq U$, described as follows:

$$G_{g_1} =< S_{1,t(1,g_1)}, S_{2,t(2,g_1)}, ..., S_{K,t(K,g_1)} >$$

$$G_{g_2} =< S_{1,t(1,g_2)}, S_{2,t(2,g_2)}, ..., S_{K,t(K,g_2)} >,$$

for $S_{j,t(j,g_1)}, S_{j,t(j,g_2)} \in V_{a_j}^m(L_j), \quad j \in \{1, 2, ..., K\}$, we can define *the group perturbations* as well as the corresponding measures. The group g_1 considered contains the objects

$$\{e_n : n \in J_{g_1} \subseteq \{1, 2, ...N\}\},$$

while the group g_2 contains the objects

$$\{e_n : n \in J_{g_2} \subseteq \{1, 2, ...N\}\},$$

where $J_{g_1} \cap J_{g_2} = \emptyset$.

Definition 5.9 (*Perturbation of group of objects*) The perturbation of one group of objects g_2 by another group of objects g_1, denoted $(G_{g_1} \mapsto G_{g_2})$, can be represented by an ordered collection of the multisets

$$(S_{j,t(j,g_1)} \mapsto S_{j,t(j,g_2)}) = S_{j,t(j,g_1)} \ominus S_{j,t(j,g_2)},$$

$j = 1, 2, ..., K$, drawn from the ordinary sets of nominal values V_{a_j} of the attributes a_j, and is defined as follows

$$(G_{g_1} \mapsto G_{g_2}) =$$

$$= < (S_{1,t(1,g_1)} \mapsto S_{1,t(1,g_2)}), ..., (S_{K,t(K,g_1)} \mapsto S_{K,t(K,g_2)}) > \quad (5.34)$$

Thus, the perturbation of one group of objects by another group of objects is defined in an analogous way to the perturbation of one object by another object. Namely, the perturbation of the one group of objects g_2 by another group of objects g_1 is represented by a collection of perturbations $S_{j,t(j,g_1)} \Theta S_{j,t(j,g_2)}$ generated for separate attributes a_j, $j = 1, 2, ..., K$. In effect, it constitutes a collection of multisets.

The counterpart case is defined analogously, that is

$$(G_{g_2} \mapsto G_{g_1}) =$$

$$= < (S_{1,t(1,g_2)} \mapsto S_{1,t(1,g_1)}), ..., (S_{K,t(K,g_2)} \mapsto S_{K,t(K,g_1)}) > \quad (5.35)$$

The measure of perturbation of one group of objects by another group of objects is a number ranging between 0 and 1 and obtained via some aggregation operation. The aggregation is done on a set of measures of the perturbations associated with each attribute a_j, for $j = 1, 2, ..., K$, see Definition 5.10.

Definition 5.10 (*Measure of perturbation of group of objects*) The measure of perturbation of the group of objects g_2 by the group of objects g_1, is denoted by $Per_{GO}(G_{g_1} \mapsto G_{g_2})$, and is defined in the following manner:

$$Per_{GO}(G_{g_1} \mapsto G_{g_2}) =$$

$$= Agg\left(Per_{MS}(S_{1,t(1,g_1)} \Theta S_{1,t(1,g_2)}), ..., Per_{MS}(S_{K,t(K,g_1)} \Theta S_{K,t(K,g_2)})\right), \quad (5.36)$$

where Agg is the aggregation operation, defined as $Agg : [0, 1]^K \to [0, 1]$.

5.6 Illustrative Examples

In order to bring closer the idea as to how to represent objects using the multisets and how the perturbations are realized, let us discus the following illustrative example.

5.6.1 Exemplary Description of Students' Semester Grades

The example concerns on the question, how to describe the object which exists in several versions (e.g., a student described by several sets of semester grades). The example was inspired by the paper of Petrovsky [28], we wanted to show the way of describing such object as a collection of multisets.

Let us consider a student e_1 and his two sets of semester grades obtained in two groups of subjects, namely obligatory and optional. Let us assume that the first set contains four obligatory subjects (attributes)

$$\{a_1, a_2, a_3, a_4\},$$

and the second set contains also four optional subject (attributes)

$$\{a_5, a_6, a_7, a_8\}.$$

All subjects have the same qualitative scale

$$V = \{v_2, v_3, v_4, v_5\} =$$

$$= \{2 - \text{“}unsatisfactory\text{”}, 3 - \text{“}satisfactory\text{”}, 4 - \text{“}good\text{”}, 5 - \text{“}excellent\text{”}\}.$$

Thus, the student e_1 (i.e., the object) is already described by two vectors of grades (i.e., values of attributes). For example, two versions of the semester grades of the student e_1, denoted by $e_1^{(1)}$ and $e_1^{(2)}$, can be represented as follows

$$e_1^{(1)} = \{(a_1 = 4), (a_2 = 5), (a_3 = 4), (a_4 = 5), (a_5 = 4), \\ (a_6 = 5), (a_7 = 4), (a_8 = 4)\},$$

$$e_1^{(2)} = \{(a_1 = 5), (a_2 = 5), (a_3 = 5), (a_4 = 5), (a_5 = 5), \\ (a_7 = 4), \ (a_8 = 4)\},$$

where the superscript (i), for $i = 1, 2$, determines the number of the semester.

By applying the multisets, the student's grades can be described in the form of one or two multisets. The use of one multiset can be found in the paper [28]. Here, we consider two multisets related to two sets of attributes, namely $\{a_1, a_2, a_3, a_4\}$ and $\{a_5, a_6, a_7, a_8\}$. The numbers of the elements are equal to number of distinguished values of the qualitative scale $V = \{v_2, v_3, v_4, v_5\}$, while each multiplicity is equal to the number of the assessments, as shown below

$$G_{e_1^{(1)}} =< S_{1,t(1,e_1^{(1)})}, S_{2,t(2,e_1^{(1)})} >=$$

$$=< \{(0, v_2), (0, v_3), (2, v_4), (2, v_5)\}, \ \{(0, v_2), (0, v_3), (3, v_4), (1, v_5)\} >,$$

$$G_{e_1^{(2)}} =< S_{1,t(1,e_1^{(2)})}, S_{2,t(2,e_1^{(2)})} >=$$

$$=< \{(0, v_2), (0, v_3), (0, v_4), (4, v_5)\}, \ \{(0, v_2), (0, v_3), (2, v_4), (1, v_5)\} > .$$

So, according to Eq. 5.23, the description of the semester grades G_{e_1} of the student e_1 is formed on the basis of two versions $G_{e_1^{(1)}}$ and $G_{e_1^{(2)}}$, and now is represented by two multisets, as shown below

$$G_{e_1} = G_{e_1^{(1)}} \oplus G_{e_1^{(2)}} =< S_{1,t(1,e_1)}, S_{2,t(2,e_1)} >=$$

$$=< \{(0, v_2), (0, v_3), (2, v_4), (6, v_5)\}, \{(0, v_2), (0, v_3), (5, v_4), (2, v_5)\} > .$$

In a similar way we can determine the description of the semester grades of another exemplary student e_2 as two other multisets, as shown below

$$G_{e_2} =< \{(1, v_2), (6, v_3), (1, v_4), (0, v_5)\}, \{(0, v_2), (4, v_3), (1, v_4), (0, v_5)\} > .$$

To the above, we consider two exemplary students e_1, e_2 represented by two descriptions G_{e_1} and G_{e_2}. Each description is represented by two multisets drawn from the ordinary sets of values $V = \{v_2, v_3, v_4, v_5\}$. According to (5.24) and (5.25), in this example, for $K = 2$, the considered multiset perturbations have the following form:

$$(G_{e_1} \mapsto G_{e_2}) =< (S_{1,t(1,e_1)} \mapsto S_{1,t(1,e_2)}), (S_{2,t(2,e_1)} \mapsto S_{2,t(2,e_2)}) >=$$

$$=< (S_{1,t(1,e_1)} \ominus S_{1,t(1,e_2)}), (S_{2,t(2,e_1)} \ominus S_{2,t(2,e_2)}) >=$$

$$=< \{(0, v_2), (0, v_3), (1, v_4), (6, v_5)\}, \{(0, v_2), (0, v_3), (4, v_4), (2, v_5)\} >,$$

$$(G_{e_2} \mapsto G_{e_1}) =< (S_{1,t(1,e_2)} \mapsto S_{1,t(1,e_1)}), (S_{2,t(2,e_2)} \mapsto S_{2,t(2,e_1)}) >=$$

$$=< (S_{1,t(1,e_2)} \ominus S_{1,t(1,e_1)}), (S_{2,t(2,e_2)} \ominus S_{2,t(2,e_1)}) >=$$

$$=< \{(1, v_2), (6, v_3), (0, v_4), (0, v_5)\}, \{(0, v_2), (4, v_3), (0, v_4), (0, v_5)\} > .$$

Thus, we have shown that the multi-attribute objects, described by a set of repeated nominal-valued attributes, can be represented by collections of multisets.

The here considered developments can be applied in various data mining problems for multi-attribute qualitative objects, wherein the values of the attributes can be repeated. For instance, object classification is based on representing each object by multisets and on a set of elementary rules, and makes if possible to assign the objects to proper groups. Thus, in the forthcoming section, the group perturbations and their measures are applied to generate the description of the groups of objects in the form of classification rules.

5.6.2 Learning Without Supervision - Preliminary Remarks

Let us consider the set of objects $e_n \in U$. We assume that the attribute values describing the objects can be repeated and the affiliation of objects to groups is not known. After preprocessing the considered data (transforming description of each object to a multisets representation) the objects are gathered into in some sense homogeneous groups. For grouping of objects we use the method proposed by Czekanowski [30]. And the resulting problem is to generate the rules $IF...THEN...$ in order to classify a new object to the proper group. The basic steps of the methodology are depicted in Fig. 5.9.

The proposed methodology consists of three main stages:

- The first stage is to preprocess the data, meaning transforming the object representation into the proper data in the form of the multiset representation.

- The next stage is to analyze the preprocessed data and gather the objects into the distinguished groups, whereas the groups are also represented by multisets. To do that, here, we use the method proposed by Czekanowski (see [30]).

- In the final stage, the descriptions of the distinguished groups of objects in the form of classification rules are generated. Each such classification rule has the following form:

$$IF\ certain\ conditions\ are\ satisfied$$

$$THEN\ a\ given\ object\ is\ a\ member\ of\ a\ specific\ group.$$

In such a rule, the conditional part will contain the disjunction of conditions related to the subsets of the values of attributes. Here, the generation of such rules is performed on the basis of perturbations of the proper multisets, which allow to distinguish the considered group from the rest of objects, belonging to other groups. The classification rules are generated for each group separately (cf. Kacprzyk and Szkatuła [31]). Finally, the generated classification rules can be applied to classify the new objects. The classification is carried out through verification of fulfilment of the conditions in the conditional parts of the rules (cf. Szkatuła [32]).

Fig. 5.9 Scheme of the approach to creation of the classification rules

The details of the third stage are presented in the forthcoming Sect. 5.6.3. The whole developed approach is illustrated by the example of grouping text documents in Sect. 5.6.4.

5.6.3 Generation of Classification Rules

Considering for example, text documents, such as articles, books, reports, etc., for which we ignore the context and the semantics, are the objects $e_n \in U$, described by the set of repeated keywords, phrases, descriptors, etc., denoted by the set of nominal, distinct values $V = \{v_1, v_2, ..., v_L\}$.

Additional information is available, concerning the multiplicity of each value v_i, $i = 1, 2, ..., L$, for each object e_n. In this way, each object e_n (i.e., a text document) can be represented by the multiset S_{e_n} drawn from the set of values V. According to (5.23), the description of the object e_n is denoted by

$$G_{e_n} =< S_{e_n} >,$$

where the multiset $S_{e_n} \in V^m(L)$ is defined as follows:

$$S_{e_n} = \{(k_{S_{en}}(v_1), v_1), (k_{S_{en}}(v_2), v_2), ..., (k_{S_{en}}(v_L), v_L)\},$$

for $v_i \in V, i = 1, 2, ..., L$. This notation states that the keyword v_i appears $k_{S_{en}}(v_i)$ times in the multiset S_{e_n}.

Let us consider two groups of objects. In the first group $g_1 \subseteq U$, there are objects

$$\{e_n : n \in J_{g_1} \subseteq \{1, 2, ...N\}\}, \quad card(J_{g_1}) = N_1,$$

while in second group $g_2 \subseteq U$, there are objects

$$\{e_n : n \in J_{g_2} \subseteq \{1, 2, ...N\}\}, \quad card(J_{g_2}) = N_2,$$

namely, the objects that do not belong to the first but belong to the second group $g_2 \subseteq U$, where therefore, $J_{g_1} \cap J_{g_2} = \emptyset$. Additionally, it is assumed that the cardinality of each group is similar (i.e., $N_1 \approx N_2$).

The classification rule, distinguishing the objects belonging to the group g_1 can be generated in the following algorithmic way.

Step 1
The groups of objects g_1 and g_2 can be represented as multisets drawn from the same set V, for $V = \{v_1, v_2, ..., v_L\}$. According to (5.33), the descriptions of the groups g_1 and g_2, denoted by

$$G_{g_1} =< S_{g_1} > \text{ and } G_{g_2} =< S_{g_2} >,$$

respectively, can be written as follows

$$S_{g_1} = \{(k_{S_{g_1}}(v_1), v_1), (k_{S_{g_1}}(v_2), v_2), ..., (k_{S_{g_1}}(v_L), v_L)\} \overset{denoted}{=}$$

$$= \{S_{g_1,v_1}, S_{g_1,v_2}, , ..., S_{g_1,v_L}\}$$

$$S_{g_2} = \{(k_{S_{g_2}}(v_1), v_1), (k_{S_{g_2}}(v_2), v_2), ..., (k_{S_{g_2}}(v_L), v_L)\} \overset{denoted}{=}$$

$$= \{S_{g_2,v_1}, S_{g_2,v_2}, , ..., S_{g_2,v_L}\},$$

which can be rewritten as

$$G_{g_1} = \bigoplus_{n \in J_{g_1}} G_{e_n} \text{ and } G_{g_2} = \bigoplus_{n \in J_{g_2}} G_{e_n}.$$

Step 2

Separately, for each keyword $v_i \in V$, for $i = 1, 2, ..., L$, the i-th measure of the perturbation of one multiset by another multiset is constructed. Such measures of perturbations are defined according to Eq. 5.10, and are called here *the elementary measures*, taking the following form

$$Per(S_{g_1,v_i} \mapsto S_{g_2,v_i}) = \frac{k_{S_{g_1}}(v_i) - k_{S_{g_1} \cap S_{g_2}}(v_i)}{k_{S_{g_1}}(v_i) + k_{S_{g_2}}(v_i)}.$$

In this way, we consider the set of L pairs of the elementary measures of perturbation and the keywords v_i, for $i = 1, 2, ..., L$. Such pairs are denoted as $PER_{S_{g_1} \mapsto S_{g_2}}$ and are written down as follows

$$PER_{S_{g_1} \mapsto S_{g_2}} = \{(Per(S_{g_1,v_1} \mapsto S_{g_2,v_1}), v_1), ..., (Per(S_{g_1,v_L} \mapsto S_{g_2,v_L}), v_L)\} =$$

$$= \left\{ \left(\frac{k_{S_{g_1}}(v_1) - k_{S_{g_1} \cap S_{g_2}}(v_1)}{k_{S_{g_1}}(v_1) + k_{S_{g_2}}(v_1)}, v_1 \right), ..., \left(\frac{k_{S_{g_1}}(v_L) - k_{S_{g_1} \cap S_{g_2}}(v_L)}{k_{S_{g_1}}(v_L) + k_{S_{g_2}}(v_L)}, v_L \right) \right\} \tag{5.37}$$

Step 3

The set of L pairs $PER_{S_{g_1} \mapsto S_{g_2}}$ of the ith elementary measure of perturbation and the keywords v_i for $i = 1, 2, ..., L$, defined by (5.37), should be rearranged by sorting with respect to their highest values of the elementary measure of perturbation. The rearrangement creates a new permutation $(i_1, i_2, , ..., i_L)$ of $(1, 2, ..., L)$ of the pairs; in result, one obtains the following set of pairs

$$PER_{S_{g_1} \mapsto S_{g_2}} = \{(Per(S_{g_1,v_i} \mapsto S_{g_2,v_i}), v_i) \mid i = i_1, i_2, ..., i_L\} \tag{5.38}$$

where the conditions

$$Per(S_{g_1,v_{i_1}} \mapsto S_{g_2,v_{i_1}}) \geq \ldots \geq Per(S_{g_1,v_{i_L}} \mapsto S_{g_2,v_{i_L}})$$

are fulfilled.

Step 4
We can consider any real number $\alpha \in [0, 1]$ treated as a parameter, called α - *threshold*. The parameter is applied to the set of sorted pairs $PER_{S_{g_1} \mapsto S_{g_2}}$, defined by (5.38), to construct a new reduced set of pairs, denoted $PER^{\alpha}_{S_{g_1} \mapsto S_{g_2}}$. The reduction is done via consideration of those pairs, for which the values of the elementary measures are greater than or equal to the value of the threshold parameter α. The new set of pairs is written in the following way

$$PER^{\alpha}_{S_{g_1} \mapsto S_{g_2}} = \{(Per(S_{g_1,v_i} \mapsto S_{g_2,v_i}), v_i) \mid i = i_1, i_2, \ldots, i_{L_\alpha}\} \tag{5.39}$$

for which

$$Per(S_{g_1,v_i} \mapsto S_{g_2,v_i}) \geq \alpha, \quad \forall i \in \{i_1, i_2, \ldots, i_{L_\alpha}\}.$$

Step 5
Then, the set of pairs $PER^{\alpha}_{S_{g_1} \mapsto S_{g_2}}$, described by (5.39), can be used to create the set of *the one – condition elementary rules* describing the group g_1. Each such one-condition elementary rule for the group g_1, denoted by $R^{\alpha}_{g_1,v_i}$, for $i = i_1, i_2, \ldots, i_{L_\alpha}$, is defined in the following manner

$$R^{\alpha}_{g_1,v_i} : \ IF \ (considered\ value = v_i); \ q(R^{\alpha}_{g_1,v_i})$$

$$THEN\ a\ given\ object\ is\ a\ member\ of\ a\ group\ g_1 \tag{5.40}$$

where $q(R^{\alpha}_{g_1,v_i})$, for $i \in \{i_1, i_2, \ldots, i_{L_\alpha}\}$, is called *the strength coefficient* of the rule $R^{\alpha}_{g_1,v_i}$, and is expressed through the elementary measure of perturbation (5.39), that is

$$q(R^{\alpha}_{g_1,v_i}) = Per(S_{g_1,v_i} \mapsto S_{g_2,v_i}).$$

It is evident that

$$0 \leq q(R^{\alpha}_{g_1,v_i}) \leq 1, \quad \forall i \in \{i_1, i_2, \ldots, i_{L_\alpha}\}.$$

Now, we consider the overall classification rule for the group g_1, denoted by $R^{\alpha}_{g_1}$, as disjunctions ($\vee$) of the one-condition elementary rules for this group, denoted by $R^{\alpha}_{g_1,v_i}$, $\forall i \in \{i_1, i_2, \ldots, i_{L_\alpha}\}$. Thus, the classification rule for the group g_1 is described in the following way:

$$R^{\alpha}_{g_1} : \ IF \ \ R^{\alpha}_{g_1,v_{i_1}} \vee R^{\alpha}_{g_1,v_{i_2}} \vee \ldots \vee R^{\alpha}_{g_1,v_{i_{L_\alpha}}}$$

$$THEN\ \ the\ given\ object\ is\ a\ member\ of\ the\ group\ g_1. \tag{5.41}$$

According to (5.40) the classification rule for the group g_1 (5.41) has the following form

$$R^{\alpha}_{g_1}: \ IF(considered\ value = v_{i_1});\ q(R^{\alpha}_{g_1,v_{i_1}}) \vee ...$$

$$... \vee (considered\ value = v_{i_{L_\alpha}});\ q(R^{\alpha}_{g_1,v_{i_{L_\alpha}}})$$

$$THEN\ the\ given\ object\ is\ a\ member\ of\ the\ group\ g_1 \quad (5.42)$$

where $q(R^{\alpha}_{g_1,v_i})$ is the strength coefficient of the one-condition elementary rule $R^{\alpha}_{g_1,v_i}$, $i \in \{i_1, i_2, ..., i_{L_\alpha}\}$.

The above procedure shows the way to create the classification rule for one group, taking into account two existing groups. When we consider more than two groups, the procedure is run in a very similar way. Namely, when generating the classification rule for the selected group g, all other groups are considered as one group containing the objects not belonging to the group g. Then (e.g., considering the classification rule for the group g_2), the objects from the rest of the groups (i.e., g_1 and g_3, g_4, and so on) are considered as constituting one group. The classification rules are formed for each group sequentially.

The already generated classification rules (5.42) (i.e., $R^{\alpha}_{g_1}$, $R^{\alpha}_{g_2}$, and so on) can be applied to classification of a new object e. The classification is carried out through verification of fulfillment of conditions in the conditional parts of the rules. The classification is unequivocal when the only one classification rule is fulfilled. In the case of equivocal situations, when more than one of the classification rules are fulfilled, the matching degree with respect to the group is calculated [32].

For example, for a new object e and the group g_1, described by the classification rule $R^{\alpha}_{g_1}$ (5.42), the matching degree $MD(e, R^{\alpha}_{g_1})$ can be calculated in the following way:

$$MD(e, R^{\alpha}_{g_1}) = MD(e,\ R^{\alpha}_{g_1,v_{i_1}} \vee R^{\alpha}_{g_1,v_{i_2}} \vee ... \vee R^{\alpha}_{g_1,v_{i_{L_\alpha}}}) =$$

$$= Agg(MD(e,\ R^{\alpha}_{g_1,v_{i_1}}), MD(e,\ R^{\alpha}_{g_1,v_{i_2}}), ..., MD(e,\ R^{\alpha}_{g_1,v_{i_{L_\alpha}}})) \quad (5.43)$$

where

$$MD(e, R^{\alpha}_{g_1,v_i}) = \begin{cases} q(R^{\alpha}_{g_1,v_i}) & if\ rule\ R^{\alpha}_{g_1,v_i}\ is\ fulfilled\ by\ object\ e \\ 0 & otherwise \end{cases}$$

where Agg is the aggregation operation (for instance - the maximum function). The value $q(R^{\alpha}_{g_1,v_i}) \in [0, 1]$, for $i = i_1, i_2, ..., i_{L_\alpha}$, is the strength coefficient of the one-condition elementary rule R_{g_1,v_i}, according to (5.40).

The developed approach to generation of the group descriptions in the form of classification rules will be illustrated by the following example.

5.6.4 Grouping of Text Documents

A practical illustration of the proposed approach is provided here for the task of grouping of the text documents, assuming that the context and the semantics are neglected. And we assume that the affiliation of objects to the groups is not known.

Here, a textual document S is modeled as a multiset, drawn from the ordinary set of unique keywords or phrases, appearing in the text. The document S can be represented by a set of L ordered pairs (where L is the number of distinguished unique keywords and phrases) according to (5.1), that is

$$S = \{(\mathit{the\ number\ of\ occurrences} \text{ of the keyword or phrase in the text document, } \mathit{the\ keyword} \text{ or } \mathit{phrase})\}.$$

Usually, the appearing keywords and phrases can be weighted in various ways, but here, for simplicity, we assume the same importance for all keywords.

- *Data processing*

Now, let us assume, that there are ten objects, being formally, text documents, $e_n \in U$, $n = 1, 2, ..., 10$, which are described by the set of six repeated keywords from the set V, $L = 6$, namely

$$V = \{keyword\#1, keyword\#2, keyword\#3, keyword\#4, keyword\#5, keyword\#6\} \overset{denoted}{=} \{v_1, v_2, v_3, v_4, v_5, v_6\}.$$

The multiplicity of each keyword is equal to the number of repetitions of the keyword and phrase appearing within the text document $e_n, n = 1, 2, ..., 10$. Hence, each text document e_n can be represented by the multiset S_{e_n} drawn from the set of values V. Thus, the descriptions of the text documents $e_1, e_2, ..., e_{10}$ can be written in the form of multisets

$$G_{e_1} =< S_{e_1} >, \quad G_{e_2} =< S_{e_2} >, \quad ..., G_{e_{10}} =< S_{e_{10}} >,$$

as the following multisets (objects):

$$S_{e_1} = \{(3, v_1), (1, v_2), (2, v_3), (0, v_4), (0, v_5), (0, v_6)\},$$

$$S_{e_2} = \{(0, v_1), (0, v_2), (0, v_3), (1, v_4), (1, v_5), (3, v_6)\},$$

$$S_{e_3} = \{(0, v_1), (1, v_2), (0, v_3), (0, v_4), (0, v_5), (4, v_6)\},$$

$$S_{e_4} = \{(2, v_1), (0, v_2), (3, v_3), (1, v_4), (0, v_5), (1, v_6)\},$$

$$S_{e_5} = \{(0, v_1), (0, v_2), (0, v_3), (1, v_4), (1, v_5), (2, v_6)\},$$

$$S_{e_6} = \{(1, v_1), (1, v_2), (2, v_3), (0, v_4), (0, v_5), (0, v_6)\},$$

$$S_{e_7} = \{(3, v_1), (0, v_2), (3, v_3), (0, v_4), (0, v_5), (0, v_6)\},$$

$$S_{e_8} = \{(0, v_1), (1, v_2), (0, v_3), (1, v_4), (1, v_5), (4, v_6)\},$$

$$S_{e_9} = \{(3, v_1), (1, v_2), (4, v_3), (0, v_4), (0, v_5), (0, v_6)\},$$

$$S_{e_{10}} = \{(0, v_1), (0, v_2), (0, v_3), (0, v_4), (1, v_5), (4, v_6)\}.$$

Having such objects (i.e., the text documents), the task is to divide these objects into similar groups and to simultaneously determine the number of these groups.

- *Division of the set of text documents into disjoint groups*

The aim of this task is to divide the set of the considered text documents U into non-empty, disjoint groups, together containing all the considered documents.

First, in order to establish the number of groups we applied the taxonomic method proposed by Czekanowski in 1909 [30]. The so called Czekanowski's diagram is a graphic methodology for grouping of multidimensional objects, which used to be widely applied in physical anthropology, plant sociology, agricultural economics, etc. The Czekanowski method is regarded as an early, perhaps the first method of cluster analysis in the world. Obviously, Czekanowski's methodology cannot be applied in all cases, but, definitely this methodology gives very important insight into the structure of the considered data set, as well as the number of groups of the data [33]. Thus, considering a set of data characterized by the same keywords, let us form a square array with cells containing the values of the measure of distances between all pairs of objects; with all the main diagonal values equal zero.

The literature of the subject provides numerous definition of distance measures. One of them is Chebyshev's distance (5.5), given as

$$D_{Chebyshev}(S_{e_p}, S_{e_q}) = \max_{i \in \{1,2,\ldots,6\}} \mid k_{S_p}(v_i) - k_{S_q}(v_i) \mid,$$

where the multisets S_{e_p} and S_{e_q} represent the documents with the counting functions $k_{S_p}(.)$ and $k_{S_q}(.)$, respectively. The value of Chebyshev distance between any pair of text documents are shown in Table 5.5.

For better visualization of the structure of the values of Chebyshev's distances between the text documents, special graphic characters (here: the black circles of different sizes) are being used. The sizes of the black circles depend on the values of Chebyshev's distances. Czekanowski's diagram with randomly arranged objects is provided in Fig. 5.10.

Now, by applying simple swapping of rows and columns, the array can be rearranged in order to gather the closest objects within distinguished groups. The proper reordering of rows and columns in the array can be treated as an unsupervised learn-

Table 5.5 The Chebyshev distances of exemplary text documents

	S_{e_1}	S_{e_2}	S_{e_3}	S_{e_4}	S_{e_5}	S_{e_6}	S_{e_7}	S_{e_8}	S_{e_9}	$S_{e_{10}}$
S_{e_1}	0	3	4	1	3	2	1	4	2	4
S_{e_2}	3	0	1	3	1	3	3	1	4	1
S_{e_3}	4	1	0	3	2	4	4	1	4	1
S_{e_4}	1	3	3	0	3	1	1	3	1	3
S_{e_5}	3	1	2	3	0	2	3	2	4	2
S_{e_6}	2	3	4	1	2	0	2	4	2	4
S_{e_7}	1	3	4	1	3	2	0	4	1	4
S_{e_8}	4	1	1	3	2	4	4	0	4	1
S_{e_9}	2	4	4	1	4	2	1	4	0	4
$S_{e_{10}}$	4	1	1	3	2	4	4	1	4	0

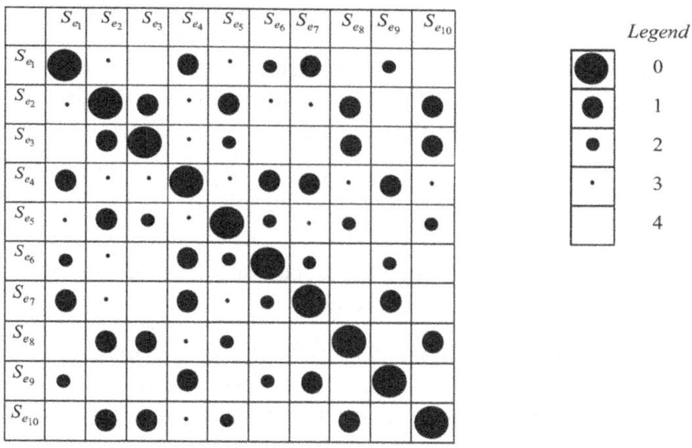

Fig. 5.10 Czekanowski's diagram with random order of exemplary objects

ing, leading to discovery of similarity as well as relationships between the objects. Formerly, in the original works by Czekanowski, the reordering of rows and columns was done manually and was very burdensome. Fortunately, nowadays, there are several computer programs for generating optimum Czekanowski's diagrams, for example the software called MaCzek [34]. In the here considered example, the transformed Czekanowski's diagram is provided in Fig. 5.11.

The rearranged objects in Fig. 5.11 clearly show that there are two distinguished groups of the considered objects, appearing as two separated blocks of used symbols. Consequently, it can be assumed that the considered text documents can be divided into two separate groups, namely

$$g_1 = \{e_1, e_4, e_6, e_7, e_9\} \text{ and } g_2 = \{e_2, e_3, e_5, e_8, e_{10}\}.$$

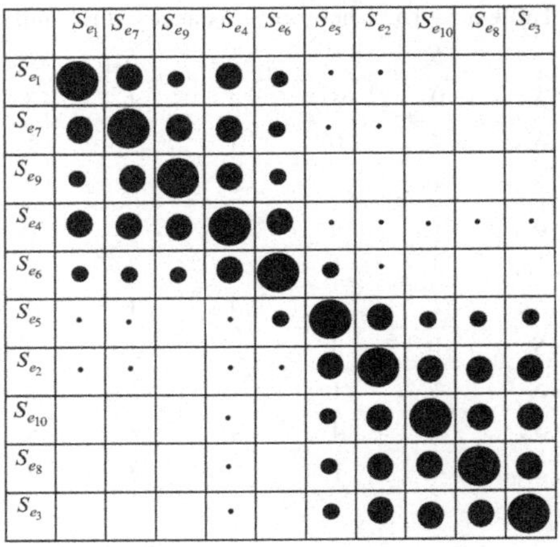

Fig. 5.11 Orderly Czekanowski diagram of exemplary objects

Then, the aim is to generate a set of elementary rules for classification of the considered texts documents into one of two groups: g_1 or g_2. Details of the applied approach can be described in the following way.

- *Generation of the classification rules*

  First, the data set is split into two sets:

  (1) *the learning set* (the first three text documents from each group),
  (2) *the testing set* (other text documents).

  Next, based on the learning set, the classification rules are generated. The testing set contains other text documents, which do not participate in the generation of the rules, and are used to check the accuracy of the classification.

  Now, let us consider the learning set, that is the group of the objects as follows

$$\{e_1, e_4, e_6\} \text{ belonging to the group } g_1, \text{ and}$$

$$\{e_2, e_3, e_5\} \text{ belonging to the group } g_2.$$

The aim is to construct the classification rule for the group g_1, as disjunctions of the one-condition elementary rules. The proper algorithm is described in the following steps.

Step 1
Let us form the descriptions of the groups g_1 and g_2 (denoted by G_{g_1} and G_{g_2}, respectively). Such descriptions are obtained by applying a simple text document aggregation. Because each object is represented by the proper multiset, then each group is also represented by the aggregated corresponding multiset. Hence, the descriptions of the considered groups g_1 and g_2 are also represented as multisets drawn from the same set V, in the following way:

$$G_{g_1} = \bigoplus_{n=1,4,6} G_{e_n} =< S_{g_1} >= \{(6, v_1), (2, v_2), (7, v_3), (1, v_4), (0, v_5), (1, v_6)\},$$

$$G_{g_2} = \bigoplus_{n=2,3,5} G_{e_n} =< S_{g_1} >= \{(0, v_1), (1, v_2), (0, v_3), (2, v_4), (2, v_5), (9, v_6)\}.$$

Step 2
Next, using the ith elementary measure of perturbation, defined as

$$Per(S_{g_1,v_i} \mapsto S_{g_2,v_i}) = \frac{k_{S_{g_1}}(v_i) - k_{S_{g_1} \cap S_{g_2}}(v_i)}{k_{S_{g_1}}(v_i) + k_{S_{g_2}}(v_i)},$$

for all $i = 1, 2, ..., 6$, we can consider the set of six following pairs, denoted by $PER_{S_{g_1} \mapsto S_{g_2}}$, due to Eq. 5.37,

$$PER_{S_{g_1} \mapsto S_{g_2}} = \{(Per(S_{g_1,v_1} \mapsto S_{g_2,v_1}), v_1), ..., (Per(S_{g_1,v_6} \mapsto S_{g_2,v_6}), v_6)\} =$$

$$= \left\{(\frac{k_{S_{g_1}}(v_1) - k_{S_{g_1} \cap S_{g_2}}(v_1)}{k_{S_{g_1}}(v_1) + k_{S_{g_2}}(v_1)}, v_1), ..., (\frac{k_{S_{g_1}}(v_6) - k_{S_{g_1} \cap S_{g_2}}(v_6)}{k_{S_{g_1}}(v_6) + k_{S_{g_2}}(v_6)}, v_6)\right\} =$$

$$= \left\{(\frac{6-0}{6+0}, v_1), (\frac{2-1}{2+1}, v_2), (\frac{7-0}{7+0}, v_3), (\frac{1-1}{3}, v_4), (\frac{0-0}{2}, v_5), (\frac{1-1}{10}, v_6)\right\} =$$

$$= \{(1.0, v_1), (0.3, v_2), (1.0, v_3), (0.0, v_4), (0.0, v_5), (0.0, v_6)\}.$$

Step 3
The above six pairs are rearranged with respect to the descending values of the elementary measures of perturbations, according to (5.38). In effect, the following set of rearranged pairs is considered:

$$PER_{S_{g_1} \mapsto S_{g_2}} = \{(1.0, v_1), (1.0, v_3), (0.3, v_2), (0.0, v_4), (0.0, v_5), (0.0, v_6)\}.$$

Step 4
Next, the value of the threshold was assumed to be $\alpha = 0.7$. Then, the reduced set of pairs, according to (5.33), for which the values of elementary measures of perturbation are greater than or equal to 0.7, has the following form:

$$PER^{0.7}_{S_{g_1} \mapsto S_{g_2}} = \{(1.0, v_1), (1.0, v_3)\}.$$

Step 5
At the final step, according to (5.42), the classification rule for the group g_1 is described as the following disjunctions of two one-condition elementary rules:

$$R^{0.7}_{g_1}: \ IF(considered\ value = v_1);\ 1.0 \ \vee \ (considered\ value = v_3);\ 1.0$$

$$THEN\ \ the\ given\ object\ is\ a\ member\ of\ the\ group\ g_1.$$

In this way the classification rule for group g_1 was generated.

Next, let us construct the classification rule for group g_2. The corresponding analogous algorithm is described below.

Step 1
Again, let us consider the descriptions of the group g_2 and g_1, denoted by G_{g_2} and G_{g_1}, respectively, in the following way:

$$G_{g_2} = \{(0, v_1), (1, v_2), (0, v_3), (2, v_4), (2, v_5), (9, v_6)\}$$

$$G_{g_1} = \{(6, v_1), (2, v_2), (7, v_3), (1, v_4), (0, v_5), (1, v_6)\}$$

Step 2
Next, using the ith elementary measure of perturbation, defined as

$$Per(S_{g2,v_i} \mapsto S_{g1,v_i}) = \frac{k_{S_{g_2}}(v_i) - k_{S_{g_2} \cap S_{g_1}}(v_i)}{k_{S_{g_2}}(v_i) + k_{S_{g_1}}(v_i)}$$

for all $i = 1, 2, ..., 6$, we can consider the set of six following pairs

$$PER_{S_{g_2} \mapsto S_{g_1}} = \{(Per(S_{g2,v_1} \mapsto S_{g1,v_1}), v_1), ..., (Per(S_{g2,v_6} \mapsto S_{g1,v_6}), v_6)\} =$$

$$= \left\{ (\frac{k_{S_{g_2}}(v_1) - k_{S_{g_2} \cap S_{g_1}}(v_1)}{k_{S_{g_2}}(v_1) + k_{S_{g_1}}(v_1)}, v_1), ..., (\frac{k_{S_{g_2}}(v_6) - k_{S_{g_2} \cap S_{g_1}}(v_6)}{k_{S_{g_2}}(v_6) + k_{S_{g_1}}(v_6)}, v_6) \right\} =$$

$$= \left\{ (\frac{0-0}{0+6}, v_1), (\frac{1-1}{1+3}, v_2), (\frac{0-0}{0+7}, v_3), (\frac{2-1}{2+1}, v_4), (\frac{2-0}{2+0}, v_5), (\frac{9-1}{9+1}, v_6) \right\} =$$

$$= \{(0.0, v_1), (0.0, v_2), (0.0, v_3), (0.3, v_4), (1.0, v_5), (0.8, v_6)\}.$$

Step 3
These six pairs are rearranged with respect to the descending values of the elementary measures of perturbations, and the following set of rearranged pairs is considered:

$$PER_{S_{g_2} \mapsto S_{g_1}} = \{(1.0, v_5), (0.8, v_6), (0.3, v_4), (0.0, v_1), (0.0, v_2), (0.0, v_3)\}.$$

Step 4
Again, the value of the threshold is assumed to be $\alpha = 0.7$, and the reduced set of pairs has the following form:

$$PER^{0.7}_{S_{g_2} \mapsto S_{g_1}} = \{(1.0, v_5), (0.8, v_6)\}.$$

Table 5.6 The classification rules for the group g_1 and g_2

	keyword#1	*keyword#3*	*keyword#5*	*keyword#6*
$R^{0.7}_{g_1}$	$q(R^{0.7}_{g_1,v_1}) = 1.0$	$q(R^{0.7}_{g_1,v_3}) = 1.0$	–	–
$R^{0.7}_{g_2}$	–	–	$q(R^{0.7}_{g_2,v_5}) = 1.0$	$q(R^{0.7}_{g_2,v_6}) = 0.8$

Step 5
At the end, the classification rule for the group g_2 is described as the following disjunction of two one-condition elementary rules:

$$R^{0.7}_{g_2}: \ IF(considered\ value = v_5);\ 1.0\ \vee\ (considered\ value = v_6);\ 0.8$$

$$THEN\ \ the\ given\ object\ is\ a\ member\ of\ the\ group\ g_2$$

In this way the classification rule for group g_2 was constructed.

- *Brief analysis of the classification rules*

Now, let us consider the generated classification rules $R^{0.7}_{g_1}$ and $R^{0.7}_{g_2}$, for the groups g_1 and g_2, respectively. Both rules are shown in Table 5.6. The number associated with each keyword is regarded as the strength coefficient of the proper elementary rule, according to (5.42).

The classification of documents to the appropriate groups is carried out through verification of fulfillment of conditions in the conditional parts of the rules.

Classification accuracy in our example is verified by applying the rules from Table 5.6 for the learning set and the testing set. Detailed calculations are presented below.

Again, let us consider the learning set, that is the six texts documents

$$\{e_1, e_4, e_6\} \text{ belonging to the group } g_1, \text{ and}$$

$$\{e_2, e_3, e_5\} \text{ belonging to the group } g_2.$$

The text documents e_1 and e_6 were unequivocally classified to the appropriate group g_1, and the text documents e_2, e_3 and e_5 were unequivocally classified to the appropriate group g_2. According to Eq. 5.43, for the text document e_4, and applying the maximum function as the aggregation operation, we obtain the following values of the matching degrees with respect to the groups g_1 and g_2

$$MD(e_4, R^{0.7}_{g_1}) = Agg(MD(e_4, R^{0.7}_{g_1,v_1}), MD(e_4, R^{0.7}_{g_1,v_3})) = Agg(1, 0) = 1,$$

$$MD(e_4, R^{0.7}_{g_2}) = Agg(MD(e_4, R^{0.7}_{g_2,v_5}), MD(e_4, R^{0.7}_{g_2,v_6})) = Agg(0.0, 0.8) = 0.8.$$

Due to the fulfillment of the inequality

$$MD(e_4, R^{0.7}_{g_1}) > MD(e_4, R^{0.7}_{g_2}),$$

the text document e_4 was correctly classified to the group g_1.

Next, let us consider the testing set (i.e., the texts documents, which have not participated in the generation of the rules),

$$\{e_7, e_9\} \text{ belonging to the group } g_1, \text{ and}$$

$$\{e_8, e_{10}\} \text{ belonging to the group } g_2.$$

Thus, for the assumptions adopted the rules also correctly classified all these texts documents.

It is worth to notice that all the considered text documents were correctly classified to the appropriate groups, according to Czekanowski's division.

The additional aim of the above described example was to illustrate the way of generating the classification rules based on Czekanowski's division as well as the developed multiset perturbation methodology.

References

1. D.E. Knuth, The art of computer programming, in *Seminumerical Algorithms*, vol. 2 (Addison-Wesley, 1969)
2. W.D. Blizard, Multiset theory. Notre Dame J. Formal Logic **30**(1), 36–66 (1989)
3. W.D. Blizard, Negative membership. Notre Dame J. Formal Logic **31**(3), 346–368 (1990)
4. K.P. Girish, S.J. John, Relations and functions in multiset context. Inf. Sci. (NY) **179**(6), 758–768 (2009)
5. V. Cerf, E. Fernandez, K. Gostelow, S. Volansky, *Formal control flow properties of a model of computation, Report UCLA-ENG-7178* (University of California, Los Angeles, CA, Computer Science Department, 1971)
6. J. Peterson, Computation sequence sets. J. Comput. Syst. Sci. **13**(1), 1–24 (1976)
7. R.R. Yager, On the theory of bags. Int. J. Gen Syst **13**, 23–37 (1986)
8. A. El-Sayed, Abo-Tabl, Topological approximations of multisets. J. Egypt. Math. Soc. **21**, 123–132 (2013)
9. K.P. Girish, J.J. Sunil, Multiset topologies induced by multiset relations. Inf. Sci. **188**, 298–313 (2012)
10. M. Krawczak, G. Szkatuła, On bilateral matching between multisets. Adv. Intell. Syst. Comput. 161–174 (2015)
11. M. Krawczak, G. Szkatuła, On perturbations of multisets, in *IEEE Symposium Series on Computational Intelligence* (South Africa, 2015), pp. 1583–1589
12. M. Krawczak, G. Szkatuła, Multiset approach to compare qualitative data, in *Proceedings 6th World Conference on Soft Computing* (Berkeley, 2016), pp. 264–269
13. A.B. Petrovsky, An axiomatic approach to metrization of multiset space, in *Multiple Criteria Decision Making*. ed. by G.H. Tzeng, H.F. Wang, U.P. Wen, P.L. Yu (Springer, New York, 1994), pp.129–1404
14. A.B. Petrovsky, Multiattribute sorting of qualitative objects in multiset spaces, in *Multiple Criteria Decision Making in the New Millennium*. Lecture Notes in Economics and Mathematical Systems, vol. 507, ed. by M. Koksalan, S. Zionts (Springer, Berlin, 2001), pp. 124–131

15. A.B. Petrovsky, Cluster analysis in multiset spaces, in *Information Systems Technology and its Applications*. ed. by M. Goldevsky, H. Mayr (Gesellschaft fur Informatik, Bonn, 2003), pp.199–206
16. D. Singh, A.M. Ibrahim, T. Yohanna, J.N. Singh, An overview of the applications of multisets. Novi Sad J. Math. **37**(2), 73–92 (2007)
17. D. Singh, A.M. Ibrahim, T. Yohanna, J.N. Singh, A systematization of fundamentals of multisets. Lecturas Matematicas **29**, 33–48 (2008)
18. A. Syropoulos, Mathematics of multisets, in *Multiset Processing*, LNCS, vol. 2235, ed. by C.S. Calude et al. (Springer, Berlin, 2001), pp. 347–358
19. R.K. Meyer, M.A. McRobbie, Multisets and relevant implication I and II. Australasian J. Philos. **60**, 107–139 and 265–281 (1982)
20. M.M. Deza, E. Deza, *Encyclopedia of Distances* (Springer, 2009), pp. 1–583
21. M. d'Amico, P. Frosini, C. Landi, Using matching distance in size theory: a survey. Int. J. Imaging Syst. Technol. **16**(5), 154–161 (2006)
22. M.M. Deza, M. Laurent, Geometry of cuts and metrics, in *Algorithms and Combinatorics*, vol. 15 (Springer, Berlin, 1997)
23. A.B. Petrovsky, Structuring techniques in multiset spaces, in *Multiple Criteria Decision Making*. Lecture Notes in Economics and Mathematical Systems, vol. 448, ed. by Fandel et al. (Springer, Berlin, 1997), pp. 174–184
24. W.A. Kosters, J.F.J. Laros, Metrics for mining multisets, in *Research and Development in Intelligent Systems XXIV, Proceedings of AI-2007*, ed. by M. Bramer, F. Coenen, M. Petridis (2007), pp. 293–303
25. C.J. Hodgetts, U. Hahn, Similarity-based asymmetries in perceptual matching. Acta Physiol. (Oxf) **139**(2), 291–299 (2012)
26. M. Krawczak, G. Szkatuła, Bidirectional comparison of multi-attribute qualitative objects. Inf. Sci. **436–437**, 367–387 (2018)
27. M. Krawczak, G. Szkatuła, An approach to dimensionality reduction in time series. Inf. Sci. **260**, 15–36 (2014)
28. A.B. Petrovsky, Methods for the group classification of multi-attribute objects (Part 1). Sci. Tech. Inf. Process. **37**(5), 346–356 (2010)
29. B. Zielman, W.J. Heiser, *Spatial Representations of Asymmetric Proximities, Research Report RR-91-10* (University of Leiden, Department of Data Theory, 1991)
30. J. Czekanowski, Zur Differentialdiagnose der Neandertalgruppe, Korespondentblatt der Deutschen Gesellschaft für Anthropologie, Ethnologie und Urgeschichte, vol. XL, No. **6**(7), 44–47 (1909)
31. J. Kacprzyk, G. Szkatuła, Inductive learning: a combinatorial optimization, in *Advances in Machine Learning I*. ed. by J. Koronacki, Z.W. Ras, S.T. Wierzchon, J. Kacprzyk (Dedicated to the Memory of Professor Ryszard S. Michalski. Springer, Heidelberg, 2010), pp.75–93
32. G. Szkatuła, Machine learning from examples under errors in data (In Polish), Ph.D. thesis (SRI PAS Warsaw, Poland, 1995)
33. I. Liiv, Seriation and matrix reordering methods: an historical overview. J. Stat. Anal. Data Mining **3**(2), 70–91 (2010)
34. A. Sołtysiak, P. Jaskulski, Czekanowski's Diagram. A method of multidimensional clustering, in *New Techniques for Old Times. CAA 98. Computer Applications and Quantitative Methods in Archaeology. Proceedings of the 26th Conference, Barcelona, March 1998*. BAR International Series, vol. 757, ed. by J.A. Barceló- I. Briz - A. Vila (Oxford, 1999), pp. 175–184

Chapter 6
Matching Between Fuzzy Sets

6.1 Preliminaries

In this chapter we consider problems, for which information is either imprecise or incomplete, or unknown. For example, description of emotions is burdened with definite subjectivity, just like the description of preferences, or judgment whether something is positive or negative. Thus, positive information states what is acceptable (e.g., what is possible or satisfactory, or permitted, or desired). Meanwhile, negative information expresses what is unacceptable (e.g., impossible or rejected, or forbidden). However, the positive and the negative information are related to each other, and they should not be taken into account separately. The classical methods, applied in modeling such kind of information usually cause either partial loss or misinterpretation of information.

The first works on the extension of two-valued logic were initiated by Jan Łukasiewicz in 1918 [1], when he proposed three-valued logic wherein the truth was denoted by 1, the falsity by 0 and the possibility by 1/2.

Later, Łukasiewicz developed the foundations of multi-valued logic [2].

There are several realizations of such multi-valued logic, and the most significant are the following.

In 1965, Lotfi Zadeh introduced the fuzzy sets theory [3]. In this theory, each element of a fuzzy set has some degree of membership, which is a real number between zero and one. The values of membership degrees were associated with the so-called membership function.

It is worth mentioning that a significant contribution to the development of fuzzy sets was made by Janusz Kacprzyk, for example, relative to the fuzzy sets in systems analysis [4], or in application to optimal control [5].

Over the years, fuzzy sets theory has developed. And so, the original fuzzy sets are now called also the fuzzy sets of type 1. While the extension of the fuzzy sets of type 1 was also invented by Zadeh in 1975, and nowadays is called fuzzy sets of type 2 (or interval fuzzy sets of type 2, or interval fuzzy sets) [6]. The essential feature of such

G. Szkatuła and M. Krawczak, *Bidirectional Comparison of Nominal Sets*,
Studies in Computational Intelligence 1140,
https://doi.org/10.1007/978-3-031-53096-8_6

fuzzy sets is related to the extension of the definition of membership functions, which can have imprecise character. Namely, for such kind of fuzzy sets two membership functions are described, the lower and the upper membership functions. In order to construct proper operations on such fuzzy sets the special arithmetics of intervals [7] has to be used.

Again, let us consider a finite, non-empty set of objects described by a set of nominal-valued attributes and, as considered in this chapter, the values of the attributes are not precisely known, and there is some information about possible or acceptable attribute values. The fuzzy sets theory allows us to model such imperfect, incomplete and inconsistent data. In the literature one finds numerous different ways of defining various forms of the fuzzy sets theory, but in all cases the membership degree lies within the range between 0 and 1.

The literature contains many measures of similarity between objects as well as corresponding formulas to calculate them (e.g., [8–13]). In the fuzzy sets theory, the value of a similarity measure of two fuzzy sets is defined in terms of membership functions, this value being obtained by a comparison of corresponding membership degrees for each element, and in result providing a value between 0 and 1.

Thus, the objects are compared each of them described by a set of inconsistent nominal-valued attributes with attributes having fuzzy sets forms. In this chapter we examine fuzzy sets and then the measure of proximity between two fuzzy sets. The considerations are based on the fuzzy sets theory and its basic operations. Notwithstanding the commonly used similarity or dissimilarity measures between two fuzzy sets, we consider a new measure of perturbation of one fuzzy set by another fuzzy set, introduced by Kacprzyk et al. [14]. After normalization, such measure of perturbation of fuzzy sets ranges from 0 to 1, where 0 is the lowest value of perturbation while 1 is the highest value of perturbation. It is shown that such measure is not always symmetrical; this means that the value of the measure of perturbation of the first fuzzy set by the second fuzzy set can be different than the value of the measure of perturbation of the second fuzzy set by the first fuzzy set. Therefore, given two measures of perturbation of two fuzzy sets we obtain more information than obtained than just the distance between the fuzzy sets.

Additionally, the sum of these measures gives valid interpretation of dissimilarity (i.e., this sum can be regarded as constitute a dissimilarity measure between two fuzzy sets). Then, we will demonstrate some special properties of the measure of perturbation between two fuzzy sets. For a better understanding of the proposed concept of the fuzzy set perturbation, the geometric interpretations are presented in two-dimensional and three-dimensional space.

6.2 Brief Introduction to Fuzzy Sets Theory

We consider a non-empty and finite set V of L nominal elements, denoted by $V = \{v_1, v_2, \ldots, v_L\}$. In 1965, Lotfi Zadeh introduced the concept of fuzzy sets as an extension of the classical sets (cf. Zadeh [3]; Kacprzyk [4]). Following Kacprzyk,

Krawczak and Szkatuła [14] let us recall the main features of the fuzzy sets theory. A fuzzy set may be represented by a collection of ordered pairs, written down in the following form:

Definition 6.1 (*Fuzzy set*) A fuzzy set B in the set V is defined in the following manner:

$$B = \{(v,\ \mu_B(v)) \mid \forall v \in V\} \tag{6.1}$$

where $\mu_B : V \to [0, 1]$ is the membership function of the fuzzy set B, that is $\mu_B(v) \in [0, 1]$ is *the degree of the membership* of the element v in the set B, $\forall v \in V$.

The membership degree expresses the degree of belongingness of the element v in expression the fuzzy set B. The expression $\mu_B(v) = 1$ means that the element v belongs for sure to the fuzzy set B (in degree 1). The expression $\mu_B(v) = 0$ means that the element v for sure does not belong to the fuzzy set B. The expression $0 < \mu_B(v) < 1$ means that the element v belongs to the fuzzy set B with the degree $\mu_B(v)$ of membership, different from "fully belonging" and "not belonging at all". The collection of all fuzzy sets in the crisp set V of nominal elements is denoted by $FS(V)$.

According to (6.1), the exemplary set $B = \{(v_1, 0.3), (v_2, 0.6), (v_3, 0.5)\}$ is a fuzzy set in the crisp set $V = \{v_1, v_2, v_3\}$.

In the fuzzy sets theory, *the negative degree of membership* (i.e., *the dissatisfaction degree*) is defined as one minus the degree of the membership, that is

$$\nu_B(v) = 1 - \mu_B(v), \quad \forall v \in V.$$

Any element of a fuzzy set (6.1) can be represented as a point having the following coordinates

$$\big(\mu_B(v),\ 1 - \mu_B(v)\big),$$

as shown in Fig. 6.1. Because the condition

$$\mu_B(v) + \nu_B(v) = 1$$

is satisfied, the whole segment, marked in bold (i.e., the segment connecting points (0,1) and (1,0)), can be treated as the image of the element v belonging to a fuzzy set V.

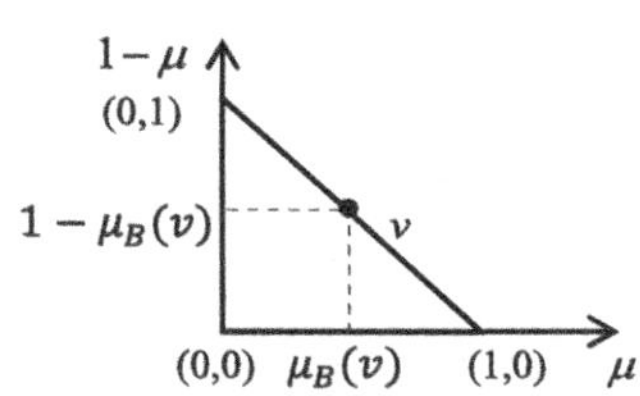

Fig. 6.1 Geometrical representation of the measures of belongingness to a fuzzy set, $v \in V$

In Fig. 6.1, the point (0,1) represents the element v definitely not belonging to the fuzzy set; the point (1,0) represents the element v belonging for sure to the fuzzy set. Any other value of belonging to the fuzzy set with some degree can be represented as a point along the segment marked in bold.

6.2.1 Operations on Fuzzy Sets

Zadeh introduced also the fundamental operations for fuzzy sets, namely the union, intersection and complementation.

Let us consider two fuzzy sets $B_i, B_j \in FS(V)$ represented by the membership functions μ_{B_i} and μ_{B_j}, respectively, where $\mu_{B_i}, \mu_{B_j} \in [0, 1], \forall v \in V$. Here, we will use the following definitions for fuzzy operations:

- $B_i \cap B_j = \{(v, \mu_{B_i \cap B_j}(v)) \mid \mu_{B_i \cap B_j}(v) := min\{\mu_{B_i}(v), \mu_{B_j}(v)\}\}$;
- $B_i \cup B_j = \{(v, \mu_{B_i \cup B_j}(v)) \mid \mu_{B_i \cup B_j}(v) := max\{\mu_{B_i}(v), \mu_{B_j}(v)\}\}$;
- $B_i \oplus B_j = \{(v, \mu_{B_i \oplus B_j}(v)) \mid \mu_{B_i \oplus B_j}(v) := \mu_{B_i}(v) + \mu_{B_j}(v)\}$;
- $B_i^C = \{(v, \mu_{B_i^C}(v)) \mid \mu_{B_i^C}(v) := 1 - \mu_{B_i}(v)\}$;
- $B_i \triangle B_j = (B_i \setminus B_j) \cup (B_j \setminus B_i) =$
  $= \{(v, \mu_{B_i \triangle B_j}(v)) \mid \mu_{B_i \triangle B_j}(v) :=\mid \mu_{B_i}(v) - \mu_{B_j}(v) \mid\}$;
- $card(B_i) = \sum_{l=1}^{L} \mu_{B_i}(v_l)$, where $L = card(V)$.

In the fuzzy space, we can consider different operations corresponding to the difference of fuzzy sets. The difference of fuzzy sets represents the features, which the first fuzzy set has but the second fuzzy set has not (e.g., Zadeh [15]).

Therefore, let us consider three sets $B_i, B_j, B_k \in FS(V)$. A function on $FS(V) \times FS(V)$ is called *a difference of fuzzy sets*, denoted by "\", when it satisfies the following conditions:

(1) if $B_i \subseteq B_j$ then $B_i \setminus B_j = \emptyset$,

(2) if $B_j \subseteq B_k$ then $(B_j \setminus B_i) \subseteq (B_k \setminus B_i)$,

$\forall B_i \in FS(V)$, and $\forall B_j \in FS(V)$ (e.g., [16]).

Here, the following three definitions of the sets difference, namely

- *the difference* between two fuzzy sets, denoted by $B_i \setminus B_j$,

$$B_i \setminus B_j = B_i \cap B_j^C, \tag{6.2}$$

• *the arithmetic subtraction*, denoted by $B_i \setminus_1 B_j$,

$$B_i \setminus_1 B_j =$$

$$= \{(v, \mu_{B_i \setminus_1 B_j}(v)) \mid \mu_{B_i \setminus_1 B_j}(v) := max\{\mu_{B_i}(v) - \mu_{B_j}(v),\ 0\}\}, \tag{6.3}$$

• *the difference operation*, denoted by $B_i \setminus_2 B_j$,

$$B_i \setminus_2 B_j = \{(v, \mu_{B_i \setminus_2 B_j}(v)) \mid \mu_{B_i \setminus_2 B_j}(v)) := \begin{cases} \mu_{B_i}(v) & \text{if } \mu_{B_j}(v) = 0 \\ 0 & \text{if } \mu_{B_j}(v) > 0 \end{cases} \tag{6.4}$$

can be considered (see Sect. 6.2.6).

A measure of fuzzy similarity or dissimilarity defines the proximity between two fuzzy sets. However, there is no unique definition of the proximity of sets and therefore we shall explore the respective axiomatic constructs.

6.2.2 Axiomatic Definitions of Measure of Similarity of Fuzzy Sets

In 1983 Wang [17] first introduced the concept of *similarity measure* of two fuzzy sets B_i and B_j in V as a function

$$S : FS(V) \times FS(V) \to [0, 1]$$

which assigns to every pair of the considered fuzzy sets a nonnegative number, with the following axiomatic conditions being satisfied:

(1s) $S(V, \emptyset) = 0$,

(2s) $\forall B_i \in FS(V),\ S(B_i, B_i) = 1$,

(3s) $\forall B_i, B_j \in FS(V),\ S(B_i, B_j) = S(B_j, B_i)$,

(4s) $\forall B_i, B_j, B_k \in FS(V)$, *if* $B_i \subseteq B_j \subseteq B_k$ then the inequalities $S(B_i, B_k) \leq S(B_i, B_j)$ and $S(B_i, B_k) \leq S(B_j, B_k)$ are satisfied.

There are many properties of a fuzzy similarity measure $S(B_i, B_j)$ between two fuzzy sets B_i and B_j in V (e.g., [8]). The fuzzy similarity measure are discussed in many works (e.g., [10, 18–20]). Below, we recall some selected properties of the fuzzy similarity measure.

A normal measure of similarity $S(B_i, B_j)$ between two fuzzy sets in V satisfies the following conditions (e.g., Liu [19]):

(1ns) $\forall B_i, B_j \in FS(V),\ S(B_i, B_j) = S(B_j, B_i)$,

(2ns) $\forall B_i \in FS(V),\ S(B_i, B_i) = 1$,

(3*ns*) $\forall B_i \in FS(V)$, $S(B_i, B_i^C) = 0$ if and only if B_i is a crisp set,

(4*ns*) $\forall B_i, B_j, B_k \in FS(V)$, if the conditions $B_i \subseteq B_j \subseteq B_k$ are satisfied, then $S(B_i, B_j) \geq S(B_i, B_k)$ and $S(B_j, B_k) \geq S(B_i, B_k)$.

On the other hand, in the paper by Raha et al. [20], the following properties of the measure of similarity between two fuzzy sets are given:

(1*ss*) $\forall B_i, B_j \in FS(V)$, $S(B_i, B_j) = S(B_j, B_i)$,

(2*ss*) $\forall B_i, B_j \in FS(V)$, $0 \leq S(B_i, B_j) \leq 1$,

(3*ss*) $\forall B_i, B_j \in FS(V)$, $S(B_i, B_j) = 1$ if and only if $B_i = B_j$,

(4*ss*) $\forall B_i, B_j, B_k \in FS(V)$, if the conditions $B_i \subseteq B_j \subseteq B_k$ are satisfied, then $S(B_i, B_k) \leq min\{S(B_i, B_j), S(B_j, B_k)\}$,

(5*ss*) ($\forall B_i, B_j \in FS(V)$, $B_i \neq \emptyset$ and $B_j \neq \emptyset$), if $S(B_i, B_j) = 0$ then $min\{\mu_{B_i}(v), \mu_{B_j}(v)\} = 0$, $\forall v \in V$.

Next, let us consider two fuzzy sets B_i and B_j on the set V, and present the most important properties of the normalized measure of similarity of two fuzzy sets.

Corollary 6.1 *The following properties of the normalized measure of similarity are true,* $\forall B_i, B_j \in FS(V)$*:*

(1) $S(B_i, B_j) = S(B_j, B_i)$,

(2) $S(B_i, B_j) = 1$ if and only if $B_i = B_j$,

(3) $S(B_i, B_j) = 0$ if and only if $B_i \cap B_j = \emptyset$,

(4) $S(B_i, B_i^C) = 1$ if and only if $\forall v \in V$, $\mu_{B_i}(v) = 0.5$,

(5) $S(B_i, B_i^C) = 0$ if and only if $\forall v \in V$, $\mu_{B_i}(v) = 1$ or $\mu_{B_i}(v) = 0$.

Thus, the similarity measures between two fuzzy sets B_i and B_j are calculated by using values of their membership functions $\mu_{B_i}(v)$ and $\mu_{B_j}(v)$, for every member of the set V, and so it has been generally assumed that the similarity is typically symmetric.

6.2.3 Examples of Measures of Similarity of Fuzzy Sets

In the literature, many, often different, measures of similarity, as well as the formulas to calculate them are proposed. Several authors have proposed similarity measures for fuzzy sets that can be viewed as generalizations of the classical set similarity measures. Hence, we would like to present some selected similarity measures already defined in the literature (e.g., [8, 9]).

In general, we can distinguish three main groups of definitions of similarity measures. Within the first group, similarity measures are based on the basic operations on fuzzy sets, such as union, intersection and cardinality, implemented as maximum (max) or/and minimum (min). The second group is based on symmetric difference operations on fuzzy sets, and the third group - on distance measures.

Such measures are recalled below (see, e.g., [10–12]).

$$S_1(B_i, B_j) = \frac{card(B_i \cap B_j)}{card(B_i \cup B_j)} = \frac{\sum_{l=1}^{L} min\{\mu_{B_i}(v_l), \mu_{B_j}(v_l)\}}{\sum_{l=1}^{L} max\{\mu_{B_i}(v_l), \mu_{B_j}(v_l)\}} \text{ for } B_i \cup B_j \neq \emptyset,$$

$$\text{if } B_i, B_j = \emptyset \text{ then } S_1(B_i, B_j) = 1;$$

$$S_2(B_i, B_j) = \sum_{l=1}^{L} \frac{min\{\mu_{B_i}(v_l), \mu_{B_j}(v_l)\}}{max\{\mu_{B_i}(v_l), \mu_{B_j}(v_l)\}} \text{ for } B_i \cup B_j \neq \emptyset,$$

$$\text{if } B_i, B_j = \emptyset \text{ then } S_2(B_i, B_j) = 1;$$

$$S_3(B_i, B_j) = \frac{card(B_i^C \cap B_j^C)}{card(B_i^C \cup B_j^C)} = \frac{\sum_{l=1}^{L} min\{(1 - \mu_{B_i}(v_l)), (1 - \mu_{B_j}(v_l))\}}{\sum_{l=1}^{L} max\{(1 - \mu_{B_i}(v_l)), (1 - \mu_{B_j}(v_l))\}};$$

$$S_4(B_i, B_j) = \frac{card(B_i \cap B_j)}{max\{card(B_i), card(B_j)\}} = \frac{\sum_{l=1}^{L} min\{\mu_{B_i}(v_l), \mu_{B_j}(v_l)\}}{max\{\sum_{l=1}^{L} \mu_{B_i}(v_l), \sum_{l=1}^{L} \mu_{B_i}(v_l)\}};$$

$$S_5(B_i, B_j) = \frac{card(B_i^C \cap B_j^C)}{max\{card(B_i^C), card(B_j^C)\}} =$$

$$= \frac{\sum_{l=1}^{L} min\{(1 - \mu_{B_i}(v_l)), (1 - \mu_{B_j}(v_l))\}}{max\{\sum_{l=1}^{L}(1 - \mu_{B_i}(v_l)), \sum_{l=1}^{L}(1 - \mu_{B_i}(v_l))\}};$$

$$S_6(B_i, B_j) = \frac{card(B_i \cap B_j)}{min\{card(B_i), card(B_j)\}} = \frac{\sum_{l=1}^{L} min\{\mu_{B_i}(v_l), \mu_{B_j}(v_l)\}}{min\{\sum_{l=1}^{L} \mu_{B_i}(v_l), \sum_{l=1}^{L} \mu_{B_i}(v_l)\}};$$

$$S_7(B_i, B_j) = S_6(B_i^C, B_j^C) = \frac{card(B_i^C \cap B_j^C)}{min\{card(B_i^C), card(B_j^C)\}} =$$

$$= \frac{\sum_{l=1}^{L} min\{(1-\mu_{B_i}(v_l)),\ (1-\mu_{B_j}(v_l))\}}{min\{\sum_{l=1}^{L}(1-\mu_{B_i}(v_l)),\ \sum_{l=1}^{L}(1-\mu_{B_i}(v_l))\}};$$

$$S_8(B_i, B_j) = \frac{1}{L}\sum_{l=1}^{L}\frac{2 \cdot min\{\mu_{B_i}(v_l),\ \mu_{B_j}(v_l)\}}{\mu_{B_i}(v_l) + \mu_{B_j}(v_l)} \quad \text{for } B_i \cup B_j \neq \emptyset,$$

$$\text{if } B_i, B_j = \emptyset \text{ then } S_8(B_i, B_j) = 1;$$

$$S_9(B_i, B_j) = \max_{l=1,\dots,L}\{min\{\mu_{B_i}(v_l),\ \mu_{B_j}(v_l)\}\};$$

$$S_{10}(B_i, B_j) = \frac{card((B_i \triangle B_j)^C)}{max\{card((B_i \triangle B_j)^C),\ card((B_j \triangle B_i)^C)\}};$$

$$S_{11}(B_i, B_j) = \frac{card((B_i \triangle B_j)^C)}{min\{card((B_i \triangle B_j)^C),\ card((B_j \triangle B_i)^C)\}};$$

$$S_{12}(B_i, B_j) = 1 - D_{12}(B_i, B_j) = 1 - \frac{card(B_i \triangle B_j)}{card(B_i \oplus B_j)} =$$

$$= 1 - \frac{\sum_{l=1}^{L} \mid \mu_{B_i}(v_l) - \mu_{B_j}(v_l) \mid}{\sum_{l=1}^{L}(\mu_{B_i}(v_l) + \mu_{B_i}(v_l))};$$

$$S_{13}(B_i, B_j) = 1 - D_{13}(B_i, B_j) = 1 - \frac{1}{L}\sum_{l=1}^{L} \mid \mu_{B_i}(v_l) - \mu_{B_j}(v_l) \mid;$$

$$S_{14}(B_i, B_j) = 1 - D_{14}(B_i, B_j) = 1 - \max_{l=1,\dots,L} \mid \mu_{B_i}(v_l) - \mu_{B_j}(v_l) \mid .$$

6.2.4 Axiomatic Definitions of Measures of Dissimilarity of Fuzzy Sets

The distance measure between two fuzzy sets can also be considered as a measure of dissimilarity between two fuzzy sets, given as a function

$$D : FS(V) \times FS(V) \rightarrow [0, 1]$$

that satisfies certain properties (axioms). In the paper by Fan and Xie [21], the following axiomatic conditions of *the measure of distance* between two fuzzy sets is given:

(1*d*) $\forall B_i, B_j \in FS(V),\ D(B_i, B_j) = D(B_j, B_i)$,

(2*d*) $\forall B_i \in FS(V),\ D(B_i, B_i) = 0$,

(3*d*) $\forall B_i \in FS(V),\ D(B_i, B_i^C) = 1$,

(4*d*) $\forall B_i, B_j, B_k \in FS(V)$, if the conditions $B_i \subseteq B_j \subseteq B_k$ are satisfied, then $\big(D(B_i, B_j) \leq D(B_i, B_k)$ and $D(B_j, B_k) \leq D(B_i, B_k)\big)$.

So, it has been assumed, again, that dissimilarity is typically symmetric.

6.2.5 Examples of Measures of Dissimilarity of Fuzzy Sets

One can find in the literature several measures of distance between two fuzzy sets $B_i, B_j \in FS(V)$, as well as some formulas for calculating these measures. Here, let us recall the most popular definitions (e.g., [5, 22]).

Let us assume that a measure of distance of fuzzy sets is defined by a mapping

$$D : FS(V) \times FS(V) \rightarrow R^+ \cup \{0\}.$$

- *Hamming distance* between the fuzzy sets B_i and B_j is given by

$$D_{Hamming}(B_i, B_j) = \sum_{l=1}^{L} \mid \mu_{B_i}(v_l) - \mu_{B_j}(v_l) \mid . \tag{6.5}$$

- *An extension of the Hamming distance* between the fuzzy sets B_i and B_j is given by

$$D_{2Hamming}(B_i, B_j) = \sum_{l=1}^{L} \mid \mu_{B_i}(v_l) - \mu_{B_j}(v_l) \mid^2 . \tag{6.6}$$

- *The generalization of the distance* can be considered as a one-parameter class of distance functions, with

$$D_{rHamming}(B_i, B_j) = (\sum_{l=1}^{L} \mid \mu_{B_i}(v_l) - \mu_{B_j}(v_l) \mid^r)^{\frac{1}{r}} . \tag{6.7}$$

- *Euclidean distance* is defined as follows

$$D_{Euclidean}(B_i, B_j) = \sqrt{\sum_{l=1}^{L}(\mu_{B_i}(v_l) - \mu_{B_j}(v_l))^2}. \tag{6.8}$$

Now, let us assume that a measure of distance of fuzzy sets is defined by the following mapping

$$D : FS(V) \times FS(V) \to [0, 1].$$

- *The normalized Hamming distance* between two fuzzy sets B_i and B_j, is defined by

$$D_{nHamming}(B_i, B_j) = \frac{1}{L}\sum_{l=1}^{L} \mid \mu_{B_i}(v_l) - \mu_{B_j}(v_l) \mid . \tag{6.9}$$

- *The normalized Euclidean distance* is defined as

$$D_{nEuclidean}(B_i, B_j) = \sqrt{\frac{1}{L}\sum_{l=1}^{L}(\mu_{B_i}(v_l) - \mu_{B_j}(v_l))^2}. \tag{6.10}$$

- *The supremum distance*, [23], is given by

$$D_{\infty}(B_i, B_j) = \sup_{v \in V} \mid \mu_{B_i}(v_l) - \mu_{B_j}(v_l) \mid . \tag{6.11}$$

It should be mentioned that in other sciences, like psychology, there are very interesting contributions to the description of object proximity (e.g., [24]).

6.2.6 Tversky's Model for Fuzzy Sets

Tversky extensively analyzed the issue of symmetry of relations between objects and he has given an axiomatic theory of object similarity [24]. Tversky's model is discussed in more detail in Sects. 2.2 and 3.2, and here we extend these considerations to the case of fuzzy sets. Let us consider two fuzzy sets, B_i and B_j, where

$B_i, B_j \in FS(V)$. For two sets, according to Eq. 3.6, Tversky derives the following family of similarity measures

$$S_{Tversky}(B_i, B_j; \alpha, \beta) = \frac{f(B_i \cap B_j)}{f(B_i \cap B_j) + \alpha \cdot f(B_i \setminus B_j) + \beta \cdot f(B_j \setminus B_i)}, \quad (6.12)$$

for some parameters $\alpha, \beta \geq 0$. This model is characterized by the different values of the parameters α and β and by the function $f(.)$.

It easy to notice that the measure of similarity, conform to Eq. 6.12, has a very general form. Selection of different values of the parameters leads to different types of similarity measures, proposed already in the literature. In this context, we can consider two cases:

(1) first, the set B_i can be treated as the reference set, while the set B_j is compared to the set B_i, or

(2) the set B_j is treated as the reference set, while the set B_i is compared to the set B_j.

The similarity measures in these two cases can be different (not symmetrical). Thus, assuming that the two sets, B_i and B_j, have the same status, the measures of similarities must be symmetric.

The similarity is based here on a function $f(.)$, which is called *the measure of a set*. Finite sets can be measured by the number of elements (i.e., the cardinality of a set), but, in general, they may be measured by any additive function that satisfies the following condition

$$f(B_i \cup B_j) = f(B_i) + f(B_j)$$

for the disjoint sets B_i and B_j (cf. [25]).

In our approach, we apply the measure of a fuzzy set $f(.)$ as the function $f : FS(V) \times FS(V) \rightarrow R^+ \cup \{0\}$ that assigns a non-negative real number, expressing the cardinality, to each finite fuzzy set $B_i \in FS(V)$, that is

$$f(B_i) = card(B_i) = \sum_{v \in V} \mu_{B_i}(v), \quad (6.13)$$

where $card(V) = L$.

In the fuzzy space, we can consider different operations of the difference of fuzzy sets, for example denoted by:

$$\text{“}\setminus\text{”}, \quad \text{“}\setminus_1\text{”} \text{ and } \text{“}\setminus_2\text{”},$$

which are discussed in details in Sect. 6.2.1. They will be applied to Tversky's measure of Eq. 6.12.

- *The difference of fuzzy sets*, denoted by $B_i \setminus B_j$ (see Eq. 6.2)

  The difference of fuzzy sets $B_i \setminus B_j$, where $B_i, B_j \in FS(V)$, can be written as

$$B_i \setminus B_j = B_i \cap B_j^C,$$

where the condition $\mu_{B_j^C}(v) = 1 - \mu_{B_j}(v),\ \forall v \in V$, is satisfied. So, we obtain

$$card(B_i \setminus B_j) = \sum_{v \in V} \mu_{B_i \setminus B_j}(v) = \sum_{v \in V} min\{\mu_{B_i}(v),\ 1 - \mu_{B_j}(v)\}.$$

In this way, Tversky's measure of Eq. 6.12 takes the form of the function $S_{Tversky} : FS(V) \times FS(V) \to [0, 1]$, which is defined for $V = \{v_1, v_2, \ldots, v_L\}$ as follows

$$S_{Tversky}(B_i, B_j; \alpha, \beta) = \frac{\sum_{l=1}^{L} min\{\mu_{B_i}(v_l),\ \mu_{B_j}(v_l)\}}{\sum_{l=1}^{L} (f_1 + \alpha \cdot f_2 + \beta \cdot f_3)}, \tag{6.14}$$

where

$$f_1 := min\{\mu_{B_i}(v_l),\ \mu_{B_j}(v_l)\},$$

$$f_2 := min\{\mu_{B_i}(v_l),\ 1 - \mu_{B_j}(v_l)\},$$

$$f_3 := min\{1 - \mu_{B_i}(v_l),\ \mu_{B_j}(v_l)\}.$$

- *The arithmetic subtraction of fuzzy sets*, denoted by $B_i \setminus_1 B_j$ (see Eq. 6.3)

  The arithmetic subtraction of two fuzzy sets can be written as

$$B_i \setminus_1 B_j = \{(v, \mu_{B_i \setminus_1 B_j}(v)) \mid \mu_{B_i \setminus_1 B_j}(v) := max\{\mu_{B_i}(v) - \mu_{B_j}(v),\ 0\}\}.$$

  From this operation, we obtain

$$card(B_i \setminus_1 B_j) = \sum_{v \in V} \mu_{B_i \setminus_1 B_j}(v) = \sum_{v \in V} max\{\mu_{B_i}(v) - \mu_{B_j}(v),\ 0\} =$$

$$= \sum_{v \in V} (\mu_{B_i}(v) - \mu_{B_i \cap B_j}(v)).$$

  Thus, for this case, the Tversky's measure, according to Eq. 6.12, can be rewritten for $V = \{v_1, v_2, \ldots, v_L\}$ in the following way

$$S_{Tversky}(B_i, B_j; \alpha, \beta) = \frac{\sum_{l=1}^{L} min\{\mu_{B_i}(v_l),\ \mu_{B_j}(v_l)\}}{\sum_{l=1}^{L}(f_1 + \alpha \cdot f_2 + \beta \cdot f_3)}, \tag{6.15}$$

where

$$f_1 := min\{\mu_{B_i}(v_n),\ \mu_{B_j}(v_n)\},$$

$$f_2 := max\{\mu_{B_i}(v_n) - \mu_{B_j}(v_n),\ 0\},$$

$$f_3 := max\{\mu_{B_j}(v_n) - \mu_{B_i}(v_n),\ 0\}.$$

As already mentioned, selection of different values for the parameters leads to different types of similarity measures. It is easy to notice that Tversky's similarity measure, for parameter values $\alpha = \beta = 1$ can be seen as the Jaccard extended coefficient

$$S_{Tversky}(B_i, B_j; 1, 1) = \frac{\sum_{l=1}^{L} \mu_{B_i \cap B_j}(v_l)}{\sum_{l=1}^{L}(\mu_{B_i}(v_l) + \mu_{B_j}(v_l) - \mu_{B_i \cap B_j}(v_l))}. \tag{6.16}$$

If $\alpha = \beta = \frac{1}{2}$, then Tversky's similarity measure is reduced to

$$S_{Tversky}(B_i, B_j; \frac{1}{2}, \frac{1}{2}) = \frac{2 \cdot \sum_{l=1}^{L} \mu_{B_i \cap B_j}(v_l)}{\sum_{l=1}^{L} \mu_{B_i}(v_l) + \sum_{l=1}^{L} \mu_{B_j}(v_l)}. \tag{6.17}$$

If $\alpha = 1$ and $\beta = 0$, then Tversky's similarity measure becomes

$$S_{Tversky}(B_i, B_j; 1, 0) = \frac{2 \cdot \sum_{l=1}^{L} \mu_{B_i \cap B_j}(v_l)}{\sum_{l=1}^{L} \mu_{B_i}(v_l)}. \tag{6.18}$$

If $\alpha = 0$ and $\beta = 1$, then Tversky's similarity measure again becomes

$$S_{Tversky}(B_i, B_j; 0, 1) = \frac{2 \cdot \sum_{l=1}^{L} \mu_{B_i \cap B_j}(v_l)}{\sum_{l=1}^{L} \mu_{B_j}(v_l)}. \tag{6.19}$$

- *The difference operation between fuzzy sets*, denoted by $B_i \setminus_2 B_j$ (see Eq. 6.4)

  The difference operation is defined as follows

$$\mu_{B_i \setminus_2 B_j}(v) = \begin{cases} \mu_{B_i}(v) & if \;\; \mu_{B_j}(v) = 0 \\ 0 & if \;\; \mu_{B_j}(v) > 0 \end{cases} \tag{6.20}$$

Let us assume that *the support of a fuzzy set* B_j, denoted by $Supp(B_j)$, is the set

$$Supp(B_j) = \{v \in V \mid \mu_{B_j}(v) > 0\}$$

(i.e., the membership function of the respective elements is non-zero). Thus, the membership function $\mu_{B_i \setminus_2 B_j}$ can be rewritten as

$$\mu_{B_i \setminus_2 B_j}(v) = \begin{cases} \mu_{B_i}(v) & if \;\; v \notin Supp(B_j) \\ 0 & if \;\; v \in Supp(B_j) \end{cases} \tag{6.21}$$

In this way, Tversky's measure, according to Eq. 6.12, can be rewritten for $V = \{v_1, v_2, ..., v_L\}$ in the following way

$$S_{Tversky}(B_i, B_j; \alpha, \beta) =$$

$$= \frac{\sum_{l=1}^{L} \mu_{B_i \cap B_j}(v_l)}{\sum_{v \in V} \mu_{B_i \cap B_j}(v_l) + \alpha \cdot \sum_{v \notin Supp(B_j)} \mu_{B_i}(v_l) + \beta \cdot \sum_{v \notin Supp(B_j)} \mu_{B_j}(v_l)}. \tag{6.22}$$

6.3 Basics of Perturbation of Fuzzy Sets

Let us assume that there is a collection of fuzzy sets $FS(V)$ on the set V, where V is a finite set of nominal, distinct elements, $V = \{v_1, v_2, ..., v_L\}$.

6.3.1 The Concept of Perturbation of Fuzzy Sets

Let us consider two fuzzy sets $B_i, B_j \in FS(V)$. Attaching of the first fuzzy set B_i to the second fuzzy set B_j may be treated as the perturbation of the second fuzzy set by the first fuzzy set - or, in other words, the fuzzy set B_i perturbs the fuzzy set B_j to some degree. In this sense, the concept of set perturbation may be understood as the impact on one fuzzy set B_j, exerted by another fuzzy set B_i, which is denoted by $(B_i \mapsto B_j)$, and interpreted as a new fuzzy set.

Definition 6.2 (*Perturbation of fuzzy sets*) Let us consider two fuzzy sets $B_i, B_j \in FS(V)$, for $V = \{v_1, v_2, ..., v_L\}$. The perturbation of the fuzzy set B_j by the fuzzy set B_i, denoted as $(B_i \mapsto B_j)$, is defined as the new fuzzy set, in the following manner:

$$(B_i \mapsto B_j) = \{(v, \ \mu_{B_i \mapsto B_j}(v)) \mid \forall v \in V\}, \tag{6.23}$$

where

$$\mu_{B_i \mapsto B_j}(v) := max\{\mu_{B_i}(v) - \alpha \cdot \mu_{B_j}(v), \ 0\},$$

for a parameter $\alpha \geq 0$.

In this case, the perturbation exhibits an inhibitory effect on B_j (when B_i affects it).

In the opposite case, the perturbation of the fuzzy set B_i by the fuzzy set B_j is defined in an analogous manner

$$(B_j \mapsto B_i) = \{(v, \ \mu_{B_j \mapsto B_i}(v)) \mid \forall v \in V\} \tag{6.24}$$

where

$$\mu_{B_j \mapsto B_i}(v) := max\{\mu_{B_j}(v) - \beta \cdot \mu_{B_i}(v), \ 0\}$$

for the parameter $\beta \geq 0$.

This model is characterized by the possibility of varying the values of the parameters α and β. In this sense one can have some extra flexibility of the expression of impact. Obviously, the parameters α and β may be assumed as different values in Eqs. 6.23 and 6.24. In a way, the parameters α and β provide for the correspondence with the original Tversky model.

In the further considerations in this chapter we assume that the values of the parameters are $\alpha = \beta = 1$, for simplicity.

The above fuzzy set impact is illustrated in the following example.

Example 6.1 Let us consider the following two fuzzy sets B_1 and B_2 in the set V of nominal elements, $V = \{v_1, v_2, v_3\}$,

$$B_1 = \{(v_1, 0.4), (v_2, 0.0), (v_3, 0.0)\},$$

$$B_2 = \{(v_1, 1.0), (v_2, 0.1), (v_3, 0.4)\}.$$

The perturbation of the fuzzy set B_2 by the fuzzy set B_1 is the empty fuzzy set, because the following condition

$$(B_1 \mapsto B_2) = \{(v_1, 0.0), (v_2, 0.0), (v_3, 0.0)\} = \emptyset$$

is satisfied. On the other hand, the perturbation of the fuzzy set B_1 by the fuzzy set B_2 is the following fuzzy set

$$(B_2 \mapsto B_1) = \{(v_1, 0.6), (v_2, 0.1), (v_3, 0.4)\}. \qquad \square$$

6.3.2 *Graphical Illustration of Perturbation of Fuzzy Sets in Two-Dimensional Space*

The geometrical interpretation of the proposed concept of perturbation of one fuzzy set by another fuzzy set, considered in two-dimensional space, is presented below.

Let us consider two-dimensional space, $card(V) = 2$ (i.e., $V = \{v_1, v_2\}$). In this space, let us consider two fuzzy sets $B_i, B_j \in FS(V)$, denoted by

$$B_i = \{(v_1, \mu_{B_i}(v_1)), (v_2, \mu_{B_i}(v_2))\},$$

$$B_j = \{(v_1, \mu_{B_j}(v_1)), (v_2, \mu_{B_j}(v_2))\}.$$

The fuzzy sets B_i and B_j can be represented as points, as this is illustrated in Fig. 6.2, and such points have the following coordinates

$$(\mu_{B_i}(v_1), \mu_{B_i}(v_2)) \text{ and } (\mu_{B_j}(v_1), \mu_{B_j}(v_2)),$$

respectively.

Often, for simplicity, the explicit mention of values $v_1, v_2 \in V$ can be omitted, when it does not lead to confusion, and then the points are characterized by a short notation of coordinates

$$(\mu^1_{B_i}, \mu^2_{B_i}) \text{ and } (\mu^1_{B_j}, \mu^2_{B_j}),$$

respectively.

Thus, the most characteristic points in this space can be written as follows, point (0,0) represents the values v_1 and v_2 completely not belonging to the fuzzy set; point (1,1) represents the values v_1 and v_2 completely belonging to the fuzzy set; point (1,0) represents the value v_1 fully belonging to the fuzzy set and the value v_2 completely not belonging to the fuzzy set; point (0,1) represents the value v_1 fully not belonging to the fuzzy set and the value v_2 fully belonging to the fuzzy set. The values v_1 and v_2, belonging to the fuzzy set with some degree, can be represented inside the square.

Now, let us discuss the relative positions of the fuzzy sets as points on the plane. For a fixed fuzzy set B_j there are four areas of possible location of any arbitrary fuzzy set B_i (i.e., the areas I, II, III and IV of Fig. 6.2), and for each point related to the fuzzy set B_i, lying within a specified area, some definite conditions are satisfied, with the detailed cases being shown in Fig. 6.2.

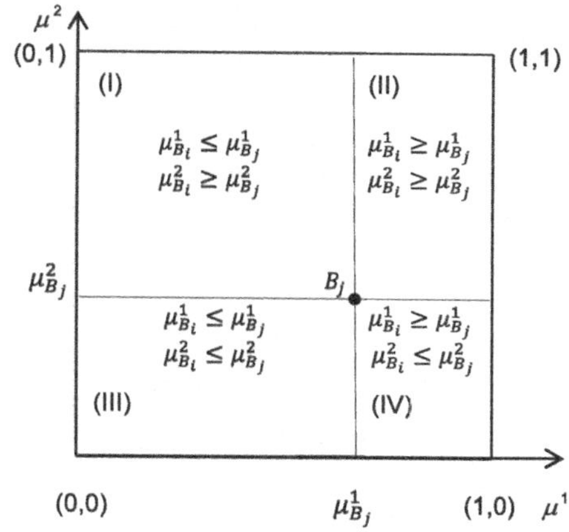

Fig. 6.2 The conditions for fixed fuzzy set B_j and arbitrary fuzzy sets B_i in four areas

According to Eq. 6.24 (for $\beta = 1$), the perturbation of an arbitrary fuzzy set B_i by another fuzzy set B_j is interpreted as a new fuzzy set, described as follows:

$$(B_j \mapsto B_i) = \{(v_1, \mu_{B_j \mapsto B_i}(v_1), \ (v_2, \mu_{B_j \mapsto B_i}(v_2)\} \tag{6.25}$$

where

$$\mu_{B_j \mapsto B_i}(v_1) = \mu_{B_j}(v_1) - \mu_{B_j \cap B_i}(v_1),$$

$$\mu_{B_j \mapsto B_i}(v_2) = \mu_{B_j}(v_2) - \mu_{B_j \cap B_i}(v_2).$$

For simplicity, the values $v_1, v_2 \in V$ can be partially omitted and the dependence, describing fuzzy set perturbation can be written down in the form

$$(B_j \mapsto B_i) = \{(v_1, \mu^1_{B_j \mapsto B_i}), \ (v_2, \mu^2_{B_j \mapsto B_i})\}.$$

In the opposite case, the perturbation of the fuzzy set B_j by the fuzzy set B_i has an analogous definition

$$(B_i \mapsto B_j) = \{(v_1, \mu_{B_i \mapsto B_j}(v_1), \ (v_2, \mu_{B_i \mapsto B_j}(v_2)\} \tag{6.26}$$

where

$$\mu_{B_i \mapsto B_j}(v_1) = \mu_{B_i}(v_1) - \mu_{B_i \cap B_j}(v_1),$$

$$\mu_{B_i \mapsto B_j}(v_2) = \mu_{B_i}(v_2) - \mu_{B_i \cap B_j}(v_2),$$

and its simplified form is as follows

$$(B_i \mapsto B_j) = \{(v_1, \mu^1_{B_i \mapsto B_j}), \ (v_2, \mu^2_{B_i \mapsto B_j})\}.$$

The geometrical interpretations of the perturbations of a fixed fuzzy set B_j and an arbitrary fuzzy set B_i are presented in Figs. 6.3, 6.4, 6.5 and 6.6. Successive figures illustrate different cases, related to the areas I, II, III or IV, respectively. Within each figure, two perturbations are indicated (i.e., $(B_j \mapsto B_i)$ and $(B_i \mapsto B_j)$). The arrows indicate the direction of the perturbation.

Let us describe and discuss the distinguished areas in details.

- *Area* I

Let us consider a fixed fuzzy set B_j, and an arbitrary fuzzy set B_i represented by any point in the area I (i.e., in the selected rectangle), as shown in Fig. 6.3.

For the fixed fuzzy set B_j, represented by the point $(\mu^1_{B_j}, \mu^2_{B_j})$, and any fuzzy set B_i, represented by a point $(\mu^1_{B_i}, \mu^2_{B_i})$ in the area I, the conditions

$$\mu^1_{B_j} \geq \mu^1_{B_i} \text{ and } \mu^2_{B_j} \leq \mu^2_{B_i}$$

are satisfied, as shown in Fig. 6.2. Due to conditions

$$\mu^1_{B_j \cap B_i} = \mu^1_{B_i} \text{ and } \mu^2_{B_j \cap B_i} = \mu^2_{B_j},$$

the perturbation $(B_j \mapsto B_i)$ is interpreted as a new fuzzy set, described as follows:

$$(B_j \mapsto B_i) = \{(v_1, \mu^1_{B_j \mapsto B_i}),\ (v_2, \mu^2_{B_j \mapsto B_i})\},$$

where

$$\mu^1_{B_j \mapsto B_i} = \mu^1_{B_j} - \mu^1_{B_j \cap B_i} = \mu^1_{B_j} - \mu^1_{B_i}$$

$$\mu^2_{B_j \mapsto B_i} = \mu^2_{B_j} - \mu^2_{B_j \cap B_i} = \mu^2_{B_j} - \mu^2_{B_j} = 0,$$

and therefore

$$(B_j \mapsto B_i) = \{(v_1, \mu^1_{B_j} - \mu^1_{B_i}),\ (v_2, 0)\}.$$

In the case of the perturbation $(B_i \mapsto B_j)$, the new fuzzy set can be written down as follows:

$$(B_i \mapsto B_j) = \{(v_1, \mu^1_{B_i \mapsto B_j}),\ (v_2, \mu^2_{B_i \mapsto B_j})\},$$

where

$$\mu^1_{B_i \mapsto B_j} = \mu^1_{B_i} - \mu^1_{B_i \cap B_j} = \mu^1_{B_i} - \mu^1_{B_i} = 0,$$

$$\mu^2_{B_i \mapsto B_j} = \mu^2_{B_i} - \mu^2_{B_i \cap B_j} = \mu^2_{B_i} - \mu^2_{B_j},$$

and therefore

$$(B_i \mapsto B_j) = \{(v_1, 0),\ (v_2, \mu^2_{B_i} - \mu^2_{B_j})\}.$$

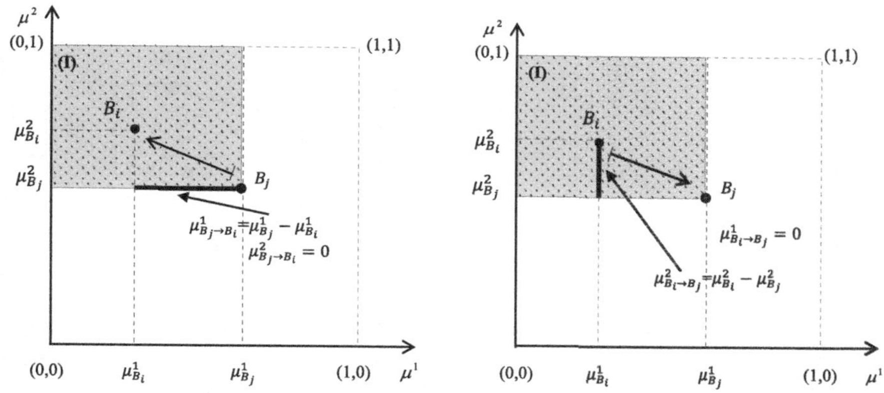

Fig. 6.3 Geometrical illustrations of the perturbations for a fixed fuzzy set B_j and an arbitrary fuzzy set B_i in the area I

A two-dimensional interpretation of the perturbations for a fixed fuzzy set B_j and an arbitrary fuzzy set B_i (according to the above formulas) is presented in Fig. 6.3.

The segments marked in bold in the figure indicate the positive values $\mu^1_{B_j \mapsto B_i}$ and $\mu^2_{B_i \mapsto B_j}$, respectively.

For the perturbation $(B_j \mapsto B_i)$, the beginning of the segment is constituted by the point $B_j = (\mu^1_{B_j}, \mu^2_{B_j})$, and the end of the segment—by the point $(\mu^1_{B_i}, \mu^2_{B_j})$.

For the opposite perturbation $(B_i \mapsto B_j)$, the beginning of the segment is constituted by the point $B_i = (\mu^1_{B_i}, \mu^2_{B_i})$, and the end of the segment—by the point $(\mu^1_{B_i}, \mu^2_{B_j})$.

- *Area* II

Let us consider a fixed fuzzy set B_j and an arbitrary fuzzy set B_i, represented by any point in the area II (i.e., in the selected rectangle), as shown in Fig. 6.4.

For the fixed fuzzy set B_j, represented by the point $(\mu^1_{B_j}, \mu^2_{B_j})$, and any fuzzy set B_i, represented by the point $(\mu^1_{B_i}, \mu^2_{B_i})$, belonging to area II, the conditions

$$\mu^1_{B_j} \le \mu^1_{B_i} \text{ and } \mu^2_{B_j} \le \mu^2_{B_i}$$

are satisfied, as shown in Fig. 6.2. Due to the conditions

$$\mu^1_{B_j \cap B_i} = \mu^1_{B_j} \text{ and } \mu^2_{B_j \cap B_i} = \mu^2_{B_j},$$

the perturbation $(B_j \mapsto B_i)$ is interpreted as a new fuzzy set, described as follows:

$$(B_j \mapsto B_i) = \{(v_1, \mu^1_{B_j \mapsto B_i}),\ (v_2, \mu^2_{B_j \mapsto B_i})\},$$

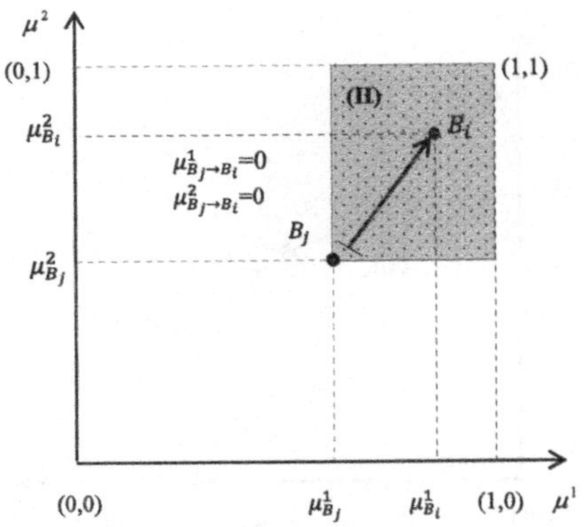

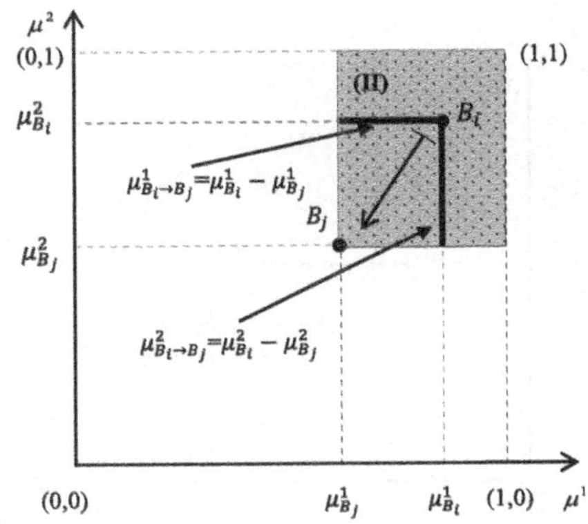

Fig. 6.4 Geometrical illustrations of the perturbations for a fixed fuzzy set B_j and an arbitrary fuzzy set B_i in the area II

where

$$\mu^1_{B_j \mapsto B_i} = \mu^1_{B_j} - \mu^1_{B_j \cap B_i} = \mu^1_{B_j} - \mu^1_{B_j} = 0,$$

$$\mu^2_{B_j \mapsto B_i} = \mu^2_{B_j} - \mu^2_{B_j \cap B_i} = \mu^2_{B_j} - \mu^2_{B_j} = 0,$$

and therefore

$$(B_j \mapsto B_i) = \{(v_1, 0),\ (v_2, 0)\}.$$

In the case of the perturbation $(B_i \mapsto B_j)$, the new fuzzy set can be written down as follows:

$$(B_i \mapsto B_j) = \{(v_1, \mu^1_{B_i \mapsto B_j}),\ (v_2, \mu^2_{B_i \mapsto B_j})\},$$

where

$$\mu^1_{B_i \mapsto B_j} = \mu^1_{B_i} - \mu^1_{B_i \cap B_j} = \mu^1_{B_i} - \mu^1_{B_j},$$

$$\mu^2_{B_i \mapsto B_j} = \mu^2_{B_i} - \mu^2_{B_i \cap B_j} = \mu^2_{B_i} - \mu^2_{B_j},$$

and therefore

$$(B_i \mapsto B_j) = \{(v_1, \mu^1_{B_i} - \mu^1_{B_j}),\ (v_2, \mu^2_{B_i} - \mu^2_{B_j})\}.$$

A two-dimensional interpretation of the perturbations for a fixed fuzzy set B_j and an arbitrary fuzzy set B_i (according to the above formulas) is presented in Fig. 6.4.

In the case of area II, the following conditions

$$\mu^1_{B_j \mapsto B_i} = 0 \text{ and } \mu^2_{B_j \mapsto B_i} = 0,$$

$$\mu^1_{B_i \mapsto B_j} = \mu^1_{B_i} - \mu^1_{B_j} \text{ and } \mu^2_{B_i \mapsto B_j} = \mu^1_{B_i} - \mu^2_{B_j},$$

are satisfied, as shown in Fig. 6.4.

It is easy to notice that for an arbitrary fuzzy set B_i, belonging to this area, the perturbation $(B_j \mapsto B_i)$ is an empty set.

For the perturbation $(B_i \mapsto B_j)$, the segments marked in bold in Fig. 6.4 indicate the positive values of $\mu^1_{B_i \mapsto B_j}$ and $\mu^2_{B_i \mapsto B_j}$, respectively. The beginning of the segments is constituted by the point $B_i = (\mu^1_{B_i}, \mu^2_{B_i})$ and the ends of the segments are the points $(\mu^1_{B_i}, \mu^2_{B_j})$ and $(\mu^1_{B_j}, \mu^2_{B_i})$, respectively.

- *Area* III

Let us consider a fixed fuzzy set B_j and an arbitrary fuzzy set B_i, represented by any point in the area III (i.e., in the selected rectangle), as shown in Fig. 6.5.

For the fixed fuzzy set B_j, represented by the point $(\mu^1_{B_j}, \mu^2_{B_j})$, and any fuzzy set B_i, represented by the point $(\mu^1_{B_i}, \mu^2_{B_i})$, belonging to area III, the conditions

$$\mu^1_{B_j} \geq \mu^1_{B_i} \text{ and } \mu^2_{B_j} \geq \mu^2_{B_i}$$

are satisfied, as shown in Fig. 6.2.

Due to conditions

$$\mu^1_{B_j \cap B_i} = \mu^1_{B_i} \text{ and } \mu^2_{B_j \cap B_i} = \mu^2_{B_i},$$

the perturbation $(B_j \mapsto B_i)$ is interpreted as a new fuzzy set, described as follows:

$$(B_j \mapsto B_i) = \{(v_1, \mu^1_{B_j \mapsto B_i}),\ (v_2, \mu^2_{B_j \mapsto B_i})\},$$

where

$$\mu^1_{B_j \mapsto B_i} = \mu^1_{B_j} - \mu^1_{B_j \cap B_i} = \mu^1_{B_j} - \mu^1_{B_i},$$

$$\mu^2_{B_j \mapsto B_i} = \mu^2_{B_j} - \mu^2_{B_j \cap B_i} = \mu^2_{B_j} - \mu^2_{B_i},$$

and therefore

$$(B_j \mapsto B_i) = \{(v_1, \mu^1_{B_j} - \mu^1_{B_i}),\ (v_2, \mu^2_{B_j} - \mu^2_{B_i})\}.$$

In the case of the perturbation $(B_i \mapsto B_j)$, the new fuzzy set can be written down as follows:

$$(B_i \mapsto B_j) = \{(v_1, \mu^1_{B_i \mapsto B_j}),\ (v_2, \mu^2_{B_i \mapsto B_j})\},$$

where

$$\mu^1_{B_i \mapsto B_j} = \mu^1_{B_i} - \mu^1_{B_i \cap B_j} = \mu^1_{B_i} - \mu^1_{B_i} = 0,$$

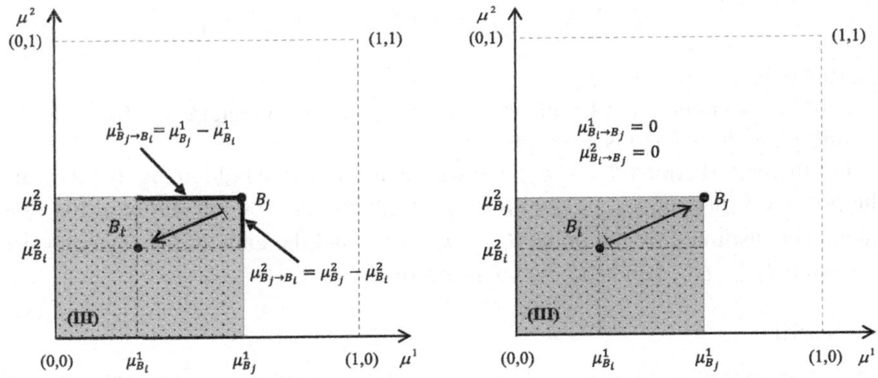

Fig. 6.5 Geometrical illustrations of the perturbations for a fixed fuzzy set B_j and an arbitrary fuzzy set B_i in the area III

$$\mu^2_{B_i \mapsto B_j} = \mu^2_{B_i} - \mu^2_{B_i \cap B_j} = \mu^2_{B_i} - \mu^2_{B_i} = 0,$$

and therefore

$$(B_i \mapsto B_j) = \{(v_1, 0),\ (v_2, 0)\}.$$

A two-dimensional interpretation of the perturbations for a fixed fuzzy set B_j and an arbitrary fuzzy set B_i (according to the above formulas) is presented in Fig. 6.5.

In the case of area III, the following conditions

$$\mu^1_{B_j \mapsto B_i} = \mu^1_{B_j} - \mu^1_{B_i} \text{ and } \mu^2_{B_j \mapsto B_i} = \mu^1_{B_j} - \mu^2_{B_i},$$

$$\mu^1_{B_i \mapsto B_j} = 0 \text{ and } \mu^2_{B_i \mapsto B_j} = 0,$$

are satisfied, as shown in Fig. 6.5.

For the perturbation $(B_j \mapsto B_i)$, the segments marked in bold in Fig. 6.5 indicate the positive values of $\mu^1_{B_j \mapsto B_i}$ and $\mu^2_{B_j \mapsto B_i}$, respectively. The beginning of the segments is constituted by the point $B_j = (\mu^1_{B_j}, \mu^2_{B_j})$, while the ends of the segments are the points $(\mu^1_{B_j}, \mu^2_{B_i})$ and $(\mu^1_{B_i}, \mu^2_{B_j})$, respectively.

Again, similarly, for an arbitrary fuzzy set B_i, belonging to this area, the perturbation $(B_i \mapsto B_j)$ is an empty set.

- *Area* IV

Let us consider a fixed fuzzy set B_j and an arbitrary fuzzy set B_i, represented by any point in the area IV (i.e., in the selected rectangle), as shown in Fig. 6.6.

For the fixed fuzzy set B_j, represented by the point $(\mu^1_{B_j}, \mu^2_{B_j})$, and any fuzzy set B_i, represented by the point $(\mu^1_{B_i}, \mu^2_{B_i})$, belonging to the area IV, the conditions

$$\mu_{B_j}^1 \leq \mu_{B_i}^1 \text{ and } \mu_{B_j}^2 \geq \mu_{B_i}^2$$

are satisfied, as shown in Fig. 6.2. Due to conditions

$$\mu_{B_j \cap B_i}^1 = \mu_{B_j}^1 \text{ and } \mu_{B_j \cap B_i}^2 = \mu_{B_i}^2,$$

the perturbation $(B_j \mapsto B_i)$ is interpreted as a new fuzzy set, described as follows:

$$(B_j \mapsto B_i) = \{(v_1, \mu_{B_j \mapsto B_i}^1), (v_2, \mu_{B_j \mapsto B_i}^2)\},$$

where

$$\mu_{B_j \mapsto B_i}^1 = \mu_{B_j}^1 - \mu_{B_j \cap B_i}^1 = \mu_{B_j}^1 - \mu_{B_j}^1 = 0$$

$$\mu_{B_j \mapsto B_i}^2 = \mu_{B_j}^2 - \mu_{B_j \cap B_i}^2 = \mu_{B_j}^2 - \mu_{B_i}^2,$$

and therefore

$$(B_j \mapsto B_i) = \{(v_1, 0), (v_2, \mu_{B_j}^2 - \mu_{B_i}^2)\}.$$

In the case of the perturbation $(B_i \mapsto B_j)$, the new fuzzy set can be written down as follows:

$$(B_i \mapsto B_j) = \{(v_1, \mu_{B_i \mapsto B_j}^1), (v_2, \mu_{B_i \mapsto B_j}^2)\},$$

where

$$\mu_{B_i \mapsto B_j}^1 = \mu_{B_i}^1 - \mu_{B_i \cap B_j}^1 = \mu_{B_i}^1 - \mu_{B_j}^1,$$

$$\mu_{B_i \mapsto B_j}^2 = \mu_{B_i}^2 - \mu_{B_i \cap B_j}^2 = \mu_{B_i}^2 - \mu_{B_i}^2 = 0,$$

and therefore

$$(B_i \mapsto B_j) = \{(v_1, \mu_{B_i}^1 - \mu_{B_j}^1), (v_2, 0)\}.$$

A two-dimensional interpretation of the perturbations for a fixed fuzzy set B_j and an arbitrary fuzzy set B_i (according to the above formulas) is presented in Fig. 6.6.

At the end, area IV is discussed, namely for an arbitrary fuzzy set $B_i \in FS(V)$, belonging to area IV, fulfillment of the following conditions

$$\mu_{B_j \mapsto B_i}^1 = 0, \quad \mu_{B_j \mapsto B_i}^2 = \mu_{B_j}^2 - \mu_{B_i}^2$$

and

$$\mu_{B_i \mapsto B_j}^1 = \mu_{B_i}^1 - \mu_{B_j}^1, \quad \mu_{B_i \mapsto B_j}^2 = 0$$

describes the cases of this area, as shown in Fig. 6.6.

The segments marked in bold in Fig. 6.6 indicate the positive values of $\mu_{B_j \mapsto B_i}^2$ and $\mu_{B_i \mapsto B_j}^1$, respectively.

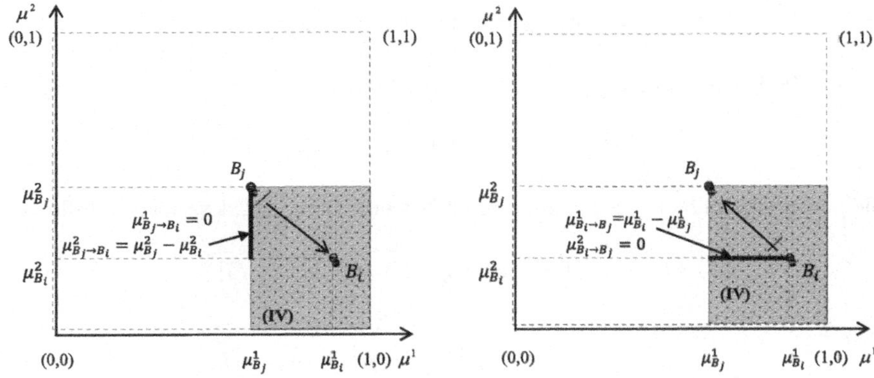

Fig. 6.6 Geometrical illustrations of the perturbations for a fixed fuzzy set B_j and an arbitrary fuzzy set B_i in the area IV

For the perturbation $(B_j \mapsto B_i)$, the beginning of the segment is constituted by the point $B_j = (\mu^1_{B_j}, \mu^2_{B_j})$, and the end of the segment - by the point $(\mu^1_{B_j}, \mu^2_{B_i})$.

For the opposite perturbation $(B_i \mapsto B_j)$, the beginning of the segment is constituted by the point $B_i = (\mu^1_{B_i}, \mu^2_{B_i})$, and the end of the segment - by the point $(\mu^1_{B_j}, \mu^2_{B_i})$.

Thus, altogether, we claim that the measure of perturbation can be understood as a measure of degree, to which two fuzzy sets match each other; however, it is important which fuzzy set is compared to which, that is, which set is considered as a prototype and which as a variant.

The here provided geometrical interpretation of the perturbations performed on fuzzy sets will be illustrated in the following example.

Example 6.2 Let us consider the set $V = \{v_1, v_2\}$ of nominal elements and two fuzzy sets $B_1, B_2 \in FS(V)$, where

$$B_1 = \{(v_1, 0.6), (v_2, 0.4)\},$$

$$B_2 = \{(v_1, 0.3), (v_2, 0.6)\}.$$

The perturbation of the fuzzy set B_2 by the fuzzy set B_1 is the following fuzzy set

$$(B_1 \mapsto B_2) = \{(v_1, \mu^1_{B_1 \mapsto B_2}),\ (v_2, \mu^2_{B_1 \mapsto B_2})\} = \{(v_1, 0.3), (v_2, 0.0)\}.$$

On the other hand, the perturbation of the fuzzy set B_1 by the fuzzy set B_2 is the following fuzzy set

$$(B_2 \mapsto B_1) = \{(v_1, \mu^1_{B_2 \mapsto B_1}),\ (v_2, \mu^2_{B_2 \mapsto B_1})\} = \{(v_1, 0.0), (v_2, 0.2)\},$$

as shown in Fig. 6.7.

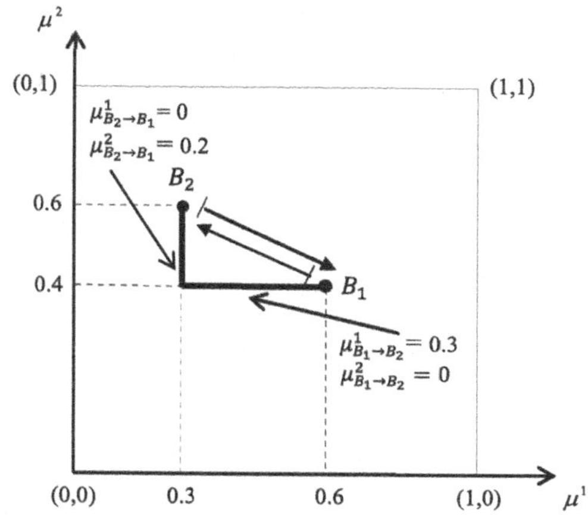

Fig. 6.7 Geometrical illustration of the perturbations for the fuzzy sets B_1 and B_2

The segments marked in bold in Fig. 6.7 indicate the positive values of $\mu^1_{B_1 \mapsto B_2}$ and $\mu^2_{B_2 \mapsto B_1}$, respectively. □

Such measures can be calculated using the values of the membership functions for every element of the set V.

6.3.3 Graphical Illustration of Perturbations of Fuzzy Sets in Three-Dimensional Space

In this section we present the graphical illustration of the concept of fuzzy sets perturbation in three-dimensional space.

Therefore, we will consider the case characterized by $card(V) = 3$ (i.e., $V = \{v_1, v_2, v_3\}$), and the fuzzy set $B_j \in FS(V)$, denoted by

$$B_j = \{(v_1, \mu_{B_j}(v_1)),\ (v_2, \mu_{B_j}(v_2)),\ (v_3, \mu_{B_j}(v_3))\},$$

where $\mu_{B_j} : V \to [0, 1]$.

The fuzzy set B_j can be represented as a point in the three-dimensional coordinate space, with coordinates

$$(\mu_{B_j}(v_1),\ \mu_{B_j}(v_2),\ \mu_{B_j}(v_3)).$$

For simplicity, the explicit mention of the elements v_1, v_2, v_3 can be omitted, and the point has the coordinates

$$(\mu^1_{B_j}, \mu^2_{B_j}, \mu^3_{B_j}),$$

see Fig. 6.8.

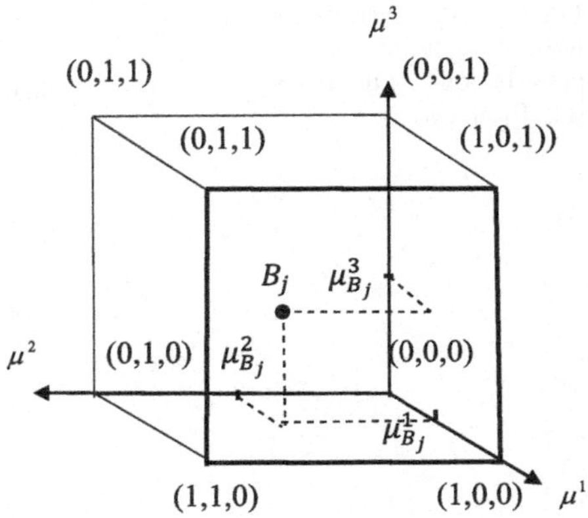

Fig. 6.8 A fuzzy set B_j, represented by the point $(\mu^1_{B_j}, \mu^2_{B_j}, \mu^3_{B_j})$, marked by •

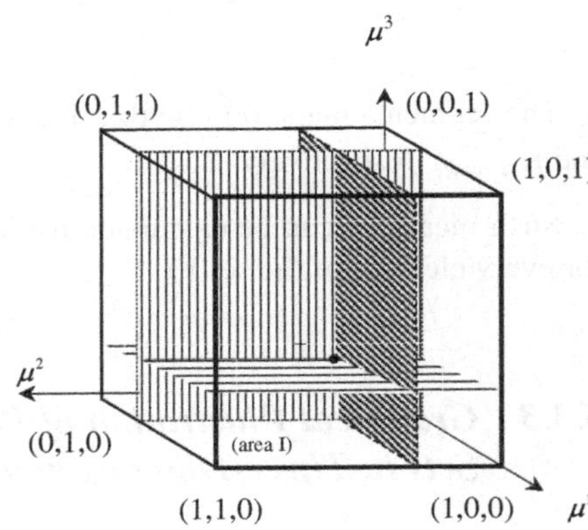

Fig. 6.9 Eight areas which are designated by the fixed fuzzy set B_j, marked by •

The most characteristic points in the three-dimensional space are the following ones, the point (0,0,0) representing the fuzzy set with the elements v_1, v_2 and v_3 completely not belonging to this fuzzy set; the point (1,1,1) representing the elements v_1, v_2 and v_3 fully belonging to the fuzzy set; the point (1,0,0) representing the element v_1 fully belonging to the fuzzy set and the elements v_2 and v_3 completely not belonging to the fuzzy set; the point (0,1,1) representing the element v_1 fully not belonging to the fuzzy set and the elements v_2 and v_3 fully belonging to the fuzzy set, etc. Any other combination of the values of belonging to the fuzzy set to some degree can be represented inside the cube, as shown in Fig. 6.8.

For the fixed fuzzy set B_j there are eight areas inside the cube (denoted I, II, ...,VIII), as shown in Fig. 6.9.

Let us consider a fuzzy set $B_i \in FS(V)$ inside the cube, denoted by

$$B_i = \{(v_1, \mu_{B_i}(v_1)), \ (v_2, \mu_{B_i}(v_2)), \ (v_3, \mu_{B_i}(v_3))\},$$

with coordinates (in the short notation)

$$(\mu^1_{B_i},\ \mu^2_{B_i},\ \mu^3_{B_i}).$$

Any point, related to the fuzzy set B_i and the point, representing fixed fuzzy set B_j satisfy some conditions. These conditions can be compared with those from Fig. 6.2 for the case of the two-dimensional space.

For example, for the fuzzy set B_i belonging to area I, which is designated by fixed fuzzy set B_j (see Fig. 6.9), the following conditions

$$\mu^1_{B_j} \le \mu^1_{B_i},\quad \mu^2_{B_j} \le \mu^2_{B_i} \text{ and } \mu^3_{B_j} \ge \mu^3_{B_i}$$

are satisfied.

According to Eq. 6.24, the perturbation of the arbitrary fuzzy set B_i by a fixed fuzzy set B_j is interpreted as a new fuzzy set, described as follows:

$$(B_j \mapsto B_i) = \{(v_1, \mu_{B_j \mapsto B_i}(v_1)),\ (v_2, \mu_{B_j \mapsto B_i}(v_2)),\ (v_3, \mu_{B_j \mapsto B_i}(v_3))\},$$

where

$$\mu_{B_j \mapsto B_i}(v) = max\{\mu_{B_j}(v) - \mu_{B_i}(v),\ 0\},\quad \forall v \in V.$$

In the short notation, the above perturbation can be expressed as follows:

$$(B_j \mapsto B_i) = \{(v_1, \mu^1_{B_j \mapsto B_i}),\ (v_2, \mu^2_{B_j \mapsto B_i}),\ (v_3, \mu^3_{B_j \mapsto B_i})\}.$$

In the case of the perturbation of the fuzzy set B_j by the fuzzy set B_i, according to Eq. 6.23, the expression is analogous

$$(B_i \mapsto B_j) = \{(v_1, \mu^1_{B_i \mapsto B_j}),\ (v_2, \mu^2_{B_i \mapsto B_j}),\ (v_3, \mu^3_{B_i \mapsto B_j})\}.$$

The geometrical interpretation of the perturbations for fixed fuzzy set B_j (represented as a point having coordinates $(\mu^1_{B_j}, \mu^2_{B_j}, \mu^3_{B_j}))$ and arbitrary fuzzy set B_i (represented as a point having coordinates $(\mu^1_{B_i}, \mu^2_{B_i}, \mu^3_{B_i})$ in the considered area) will be presented in more detail for selected areas (i.e., III and VII).

- *Area* III

Let us consider a fixed fuzzy set B_j and an arbitrary fuzzy sets B_i belonging to area III (see Fig. 6.10). The respective cube is marked by blue line.

For these two fuzzy sets the following conditions

$$\mu^1_{B_j} \ge \mu^1_{B_i},\quad \mu^2_{B_j} \ge \mu^2_{B_i} \text{ and } \mu^3_{B_j} \ge \mu^3_{B_i}$$

are satisfied.

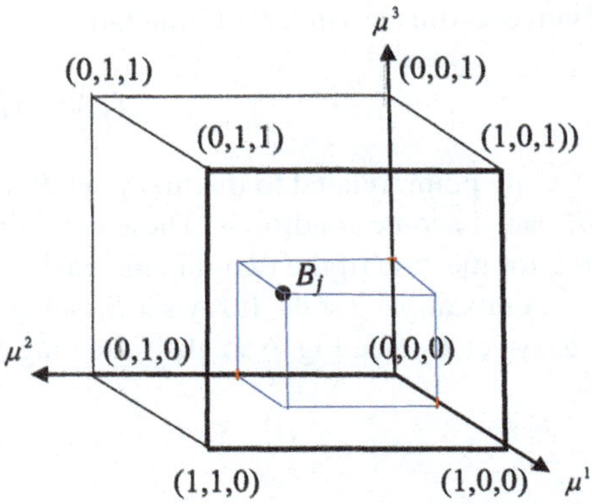

Fig. 6.10 The area III of possible location of arbitrary fuzzy sets B_i, the respective cube is marked by blue line

Let us consider the perturbation $(B_j \mapsto B_i)$. Due to conditions

$$\mu^1_{B_j \cap B_i} = \mu^1_{B_i}, \quad \mu^2_{B_j \cap B_i} = \mu^2_{B_i} \text{ and } \mu^3_{B_j \cap B_i} = \mu^3_{B_i},$$

the perturbation $(B_j \mapsto B_i)$ is interpreted as a new fuzzy set, described as follows:

$$(B_j \mapsto B_i) = \{(v_1, \mu^1_{B_j \mapsto B_i}), (v_2, \mu^2_{B_j \mapsto B_i}), (v_3, \mu^3_{B_j \mapsto B_i})\},$$

where

$$\mu^1_{B_j \mapsto B_i} = \mu^1_{B_j} - \mu^1_{B_j \cap B_i} = \mu^1_{B_j} - \mu^1_{B_i},$$

$$\mu^2_{B_j \mapsto B_i} = \mu^2_{B_j} - \mu^2_{B_j \cap B_i} = \mu^2_{B_j} - \mu^2_{B_i},$$

$$\mu^3_{B_j \mapsto B_i} = \mu^3_{B_j} - \mu^3_{B_j \cap B_i} = \mu^3_{B_j} - \mu^3_{B_i}.$$

A three-dimensional interpretation of the perturbation $(B_j \mapsto B_i)$ for a fixed fuzzy set B_j and an arbitrary fuzzy set B_i (according to the above formulas) is presented on the left hand side of Fig. 6.11. The arrow indicates the direction of the perturbation, and the segments marked in red indicate the positive values of the membership functions of the perturbation. The fragment from the left hand side of Fig. 6.11, drawn in blue lines, is enlarged and shown in details on the right hand side of the figure.

The segments marked in red indicate the positive values of

$$\mu^1_{B_j \mapsto B_i}, \quad \mu^2_{B_j \mapsto B_i} \text{ and } \mu^3_{B_j \mapsto B_i},$$

respectively. The beginning of the segments is constituted by the point

$$B_j = (\mu^1_{B_j}, \mu^2_{B_j}, \mu^3_{B_j})$$

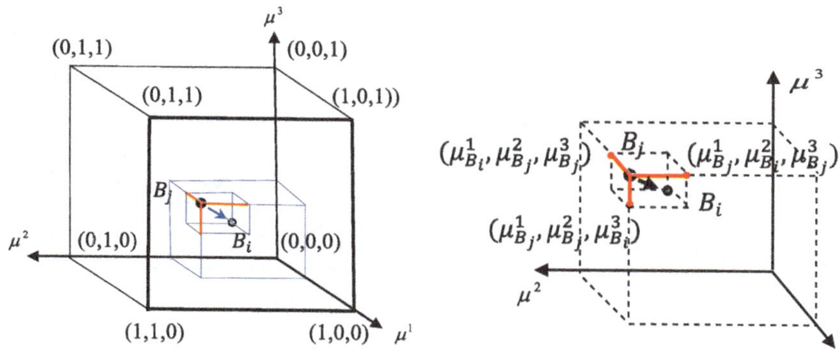

Fig. 6.11 Geometrical illustration of perturbation $(B_j \mapsto B_i)$, the fuzzy sets B_j and B_i are marked by dots • and ∘

and the ends of the segments are constituted by the points

$$(\mu^1_{B_i}, \mu^2_{B_j}, \mu^3_{B_j}),\ \ (\mu^1_{B_j}, \mu^2_{B_i}, \mu^3_{B_j})\ \text{ and }\ (\mu^1_{B_j}, \mu^2_{B_j}, \mu^3_{B_i}),$$

respectively.

Let us consider the perturbation $(B_i \mapsto B_j)$. Due to the satisfied conditions

$$\mu^1_{B_j \cap B_i} = \mu^1_{B_i},\ \ \mu^2_{B_j \cap B_i} = \mu^2_{B_i}\ \text{ and }\ \mu^3_{B_j \cap B_i} = \mu^3_{B_i},$$

the perturbation is interpreted as a new fuzzy set, described as follows:

$$(B_i \mapsto B_j) = \{(v_1, \mu^1_{B_i \mapsto B_j}),\ (v_2, \mu^2_{B_i \mapsto B_j}),\ (v_3, \mu^3_{B_i \mapsto B_j})\},$$

where

$$\mu^1_{B_i \mapsto B_j} = \mu^1_{B_i} - \mu^1_{B_i \cap B_j} = \mu^1_{B_i} - \mu^1_{B_i} = 0,$$

$$\mu^2_{B_i \mapsto B_j} = \mu^2_{B_i} - \mu^2_{B_i \cap B_j} = \mu^2_{B_i} - \mu^2_{B_i} = 0,$$

$$\mu^3_{B_i \mapsto B_j} = \mu^3_{B_i} - \mu^3_{B_i \cap B_j} = \mu^3_{B_i} - \mu^3_{B_i} = 0.$$

The three-dimensional interpretation of the perturbation $(B_i \mapsto B_j)$ for a fixed fuzzy set B_j and an arbitrary fuzzy set B_i (according to the above formulas) is presented on the left hand side of Fig. 6.12. The arrow indicates the direction of the perturbation. The right hand side of Fig. 6.12 shows the enlarged blue line space from the left hand side of Fig. 6.12.

It is easy to notice that the perturbation $(B_i \mapsto B_j)$ is an empty set, because the conditions

$$\mu^1_{B_i \mapsto B_j} = 0,\ \ \mu^2_{B_i \mapsto B_j} = 0\ \text{ and }\ \mu^3_{B_i \mapsto B_j} = 0$$

are satisfied.

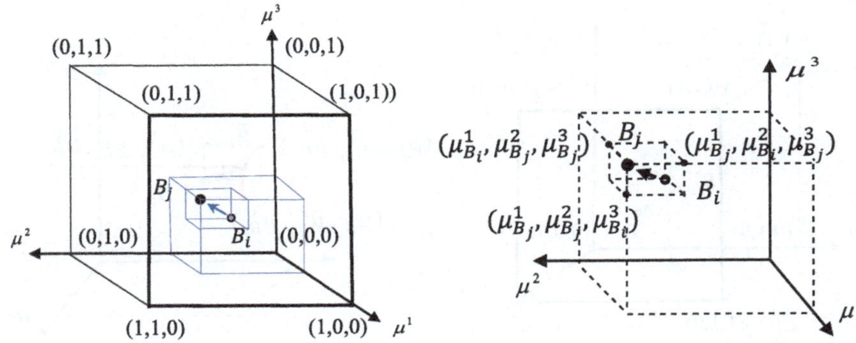

Fig. 6.12 Geometrical illustration of perturbation ($B_i \mapsto B_j$), the fuzzy sets B_i and B_j are marked by dots ∘ and •

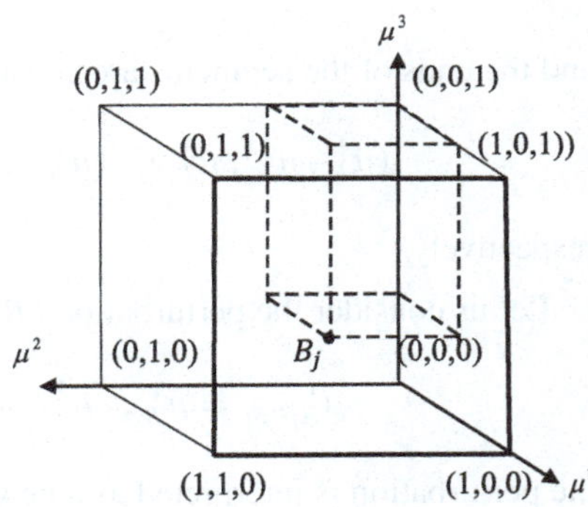

Fig. 6.13 Area VII of possible location of an arbitrary fuzzy set B_i

- *Area* VII

Let us consider a fixed fuzzy set B_j and area VII (i.e., the cube, marked by the dotted line, as shown in Fig. 6.13).

For a fixed fuzzy set B_j and an arbitrary fuzzy set B_i, belonging to area VII, the conditions

$$\mu^1_{B_j} \geq \mu^1_{B_i}, \quad \mu^2_{B_j} \geq \mu^2_{B_i} \text{ and } \mu^3_{B_j} \leq \mu^3_{B_i}$$

are satisfied.

A three-dimensional interpretation of the impact on a fixed fuzzy set B_j and an arbitrary fuzzy set B_i is presented in Fig. 6.14. In this figure, there are two perturbations (i.e., ($B_j \mapsto B_i$) and ($B_i \mapsto B_j$)). The arrow indicates the direction of the impact.

When analyzing Fig. 6.14, one may notice that for an arbitrary fuzzy set B_i, belonging to area VII, the following conditions

$$\mu^1_{B_j \mapsto B_i} = \mu^1_{B_j} - \mu^1_{B_i}, \quad \mu^2_{B_j \mapsto B_i} = \mu^2_{B_j} - \mu^2_{B_i} \text{ and } \mu^3_{B_j \mapsto B_i} = 0$$

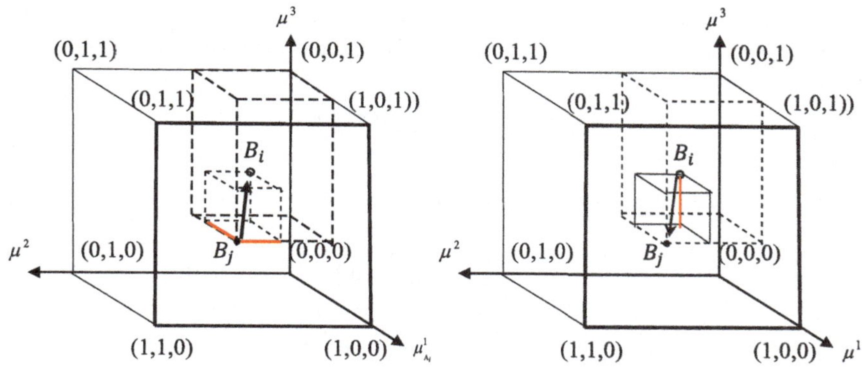

Fig. 6.14 Interpretation of the perturbations for fixed fuzzy set B_j and arbitrary fuzzy set B_i

are satisfied, and the segments marked in red indicate the positive values of $\mu^1_{B_j \mapsto B_i}$ and $\mu^2_{B_j \mapsto B_i}$, respectively (see the left hand side of Fig. 6.14).

In the opposite case, the conditions

$$\mu^1_{B_i \mapsto B_j} = 0, \quad \mu^2_{B_i \mapsto B_j} = 0 \text{ and } \mu^3_{B_i \mapsto B_j} = \mu^3_{B_i} - \mu^3_{B_j}$$

are satisfied, and the segment marked in red indicates the positive value $\mu^3_{B_i \mapsto B_j}$ (see the right hand side of Fig. 6.14).

6.3.4 *Measures of Perturbation of Fuzzy Sets*

In this section, we give a proposal of normalization of the measure of perturbation of one fuzzy set by another fuzzy set.

Definition 6.3 (*Measure of perturbation of fuzzy sets*) Assume that fuzzy sets $B_i, B_j \in FS(V)$, for $V = \{v_1, v_2, ..., v_L\}$. The measure of perturbation of the fuzzy set B_j by the fuzzy set B_i, denoted $Per_{FS}(B_i \mapsto B_j)$, is defined by the mapping $Per_{FS} : FS(V) \times FS(V) \to [0, 1]$, given as

$$Per_{FS}(B_i \mapsto B_j) = \frac{card(B_i \ominus B_j)}{card(B_i \oplus B_j)}. \tag{6.27}$$

In other words, the degree of perturbation of one fuzzy set by another fuzzy set is measured as a ratio of cardinal numbers of two fuzzy sets, one fuzzy set describing the arithmetic subtraction of the considered fuzzy sets, while the other fuzzy set corresponds to the arithmetic addition of the considered fuzzy sets.

In our approach, we apply *the cardinality of any fuzzy set* as a function that assigns to each finite fuzzy set $B_i \in FS(V)$ a non-negative real number, in the following way

$$card(B_i) = \sum_{v \in V} \mu_{B_i}(v).$$

Thus, Eq. 6.27 can be rewritten as follows

$$Per_{FS}(B_i \mapsto B_j) = \frac{\sum_{l=1}^{L} max\{\mu_{Bi}(v_l) - \mu_{Bj}(v_l),\ 0\}}{\sum_{l=1}^{L}(\mu_{Bi}(v_l) + \mu_{Bj}(v_l))} =$$

$$= \frac{\sum_{l=1}^{L}(\mu_{B_i}(v_l) - \mu_{B_i \cap B_j}(v_l))}{\sum_{l=1}^{L}(\mu_{B_i}(v_l) + \mu_{B_j}(v_l))}. \tag{6.28}$$

In the opposite case, namely that if perturbation of the fuzzy set B_i by the fuzzy set B_j, the definition of the measure is analogous, and so can be expressed as follows:

$$Per_{FS}(B_j \mapsto B_i) = \frac{\sum_{l=1}^{L} max\{\mu_{B_j}(v_l) - \mu_{B_i}(v_l),\ 0\}}{\sum_{l=1}^{L}(\mu_{B_j}(v_l) + \mu_{B_i}(v_l))} =$$

$$= \frac{\sum_{l=1}^{L}(\mu_{B_j}(v_l) - \mu_{B_j \cap B_i}(v_l))}{\sum_{l=1}^{L}(\mu_{B_j}(v_l) + \mu_{B_i}(v_l))}. \tag{6.29}$$

Let us illustrate the interpretation of the proposed concept of perturbation of one fuzzy set by another fuzzy set by the following example.

Example 6.3 Assume that we consider two fuzzy sets, B_1 and B_2, in the set $V = \{v_1, v_2, v_3\}$, where

$$B_1 = \{(v_1, 0.2), (v_2, 0.2), (v_3, 0.2)\},$$

$$B_2 = \{(v_1, 0.3), (v_2, 0.3), (v_3, 0.1)\}.$$

Using Eqs. 6.28 and 6.29 we obtain the following values of the measures of perturbation of fuzzy sets, namely:

$$Per_{FS}(B_1 \mapsto B_2) = \frac{\sum_{n=1}^{3}(\mu_{B_1}(v_n) - \mu_{B_1 \cap B_2}(v_n))}{\sum_{n=1}^{3}(\mu_{B_1}(v_n) + \mu_{B_2}(v_n))} = \frac{1}{13},$$

$$Per_{FS}(B_2 \mapsto B_1) = \frac{\sum_{n=1}^{3}(\mu_{B_2}(v_n) - \mu_{B_2 \cap B_1}(v_n))}{\sum_{n=1}^{3}(\mu_{B_2}(v_n) + \mu_{B_1}(v_n))} = \frac{2}{13},$$

which are different, meaning that this pair of perturbations is not symmetrical. □

6.3.5 Properties of Measures of Perturbation of Fuzzy Sets

Now, let us present a short review of the most important features of the proposed measure of fuzzy set perturbation. Again, let us consider two fuzzy sets, B_i and B_j, in the set V.

Corollary 6.2 *The following properties of the measures of perturbation are true:*

(1) $\forall B_i, B_j \in FS(V)$, $Per_{FS}(B_i \mapsto B_j) \in [0, 1]$;

(2) $Per_{FS}(B_i \mapsto B_j) = 1$ *if and only if* $B_j = \emptyset$ *and* $B_i \neq \emptyset$;

(3) $Per_{FS}(B_i \mapsto B_j) = 0$ *if and only if* $B_i \subseteq B_j$;

(4) $\forall B_i, B_j \in FS(V)$ *the following conditions* $0 \leq Per_{FS}(B_i \mapsto B_j) + Per_{FS}(B_j \mapsto B_i) \leq 1$ *are satisfied.*

Proof Item *(1)*

(1) Let us prove the inequality $Per_{FS}(B_i \mapsto B_j) \geq 0$. It should be noticed that inequality

$$\mu_{B_i \cap B_j}(v) \leq \mu_{B_i}(v), \quad \forall v \in V$$

is satisfied, so $\mu_{B_i}(v) - \mu_{B_i \cap B_j}(v) \geq 0$. By Eq. 6.28, we obtain inequality $Per_{FS}(B_i \mapsto B_j) \geq 0$.

(2) Now, we will consider the inequality $Per_{FS}(B_i \mapsto B_j) \leq 1$. Considering two fuzzy sets B_i and B_j in V, it should be noticed that inequality

$$\mu_{B_i}(v) - \mu_{B_i \cap B_j}(v) \leq \mu_{B_i}(v) + \mu_{B_j}(v), \quad \forall v \in V,$$

is satisfied, and then we can obtain the following inequality

$$Per_{FS}(B_i \mapsto B_j) = \frac{\sum_{l=1}^{L}(\mu_{B_i}(v_l) - \mu_{B_i \cap B_j}(v_l))}{\sum_{l=1}^{L}(\mu_{B_i}(v_l) + \mu_{B_j}(v_l))} \leq$$

$$\leq \frac{\sum_{l=1}^{L}(\mu_{B_i}(v_l) + \mu_{B_j}(v_l))}{\sum_{l=1}^{L}(\mu_{B_i}(v_l) + \mu_{B_j}(v_l))} = 1.$$

Proof Item *(2)*

(1) Let us prove the implication:

$$if \;\; Per_{FS}(B_i \mapsto B_j) = 1 \;\; then \;\; B_j = \emptyset.$$

Let us assume that $Per_{FS}(B_i \mapsto B_j) = 1$. By Eq. 6.28, the equality

$$\mu_{B_i}(v) - \mu_{B_i \cap B_j}(v) = \mu_{B_i}(v) + \mu_{B_j}(v), \;\; \forall v \in V,$$

is satisfied. It should be noticed that the equality $\mu_{B_j}(v) = -\mu_{B_i \cap B_j}(v)$ is satisfied, so $\mu_{B_j}(v) + \mu_{B_i \cap B_j}(v) = 0$, and hence we can obtain $B_j = \emptyset$.

(2) Consider now the implication:

$$if \;\; (B_j = \emptyset \;\; and \;\; B_i \neq \emptyset) \;\; then \;\; Per_{FS}(B_i \mapsto B_j) = 1.$$

Let us assume that $B_j = \emptyset$, thus $\mu_{B_i}(v) - \mu_{B_i \cap B_j}(v) = \mu_{B_i}(v)$ and $\mu_{B_i}(v) + \mu_{B_j}(v) = \mu_{B_i}(v)$, $\forall v \in V$. In this manner, we get

$$Per_{FS}(B_i \mapsto B_j) = \frac{\sum_{l=1}^{L}(\mu_{B_i}(v_l) - \mu_{B_i \cap B_j}(v_l))}{\sum_{l=1}^{L}(\mu_{B_i}(v_l) + \mu_{B_j}(v_l))} = \frac{\sum_{l=1}^{L}\mu_{B_i}(v_l)}{\sum_{l=1}^{L}\mu_{B_i}(v_l)} = 1.$$

Proof Item *(3)*

(1) Let us prove the implication:

$$if \;\; Per_{FS}(B_i \mapsto B_j) = 0 \;\; then \;\; B_i \subseteq B_j.$$

Assume that $Per_{FS}(B_i \mapsto B_j) = 0$. By Eq. 6.28, the function $Per_{FS}(B_i \mapsto B_j)$ is non-negative, and reaches the minimum if the condition

$$\mu_{B_i}(v) - \mu_{B_i \cap B_j}(v) = 0, \;\; \forall v \in V$$

is satisfied. If $\mu_{B_i}(v) = \mu_{B_i \cap B_j}(v)$, then the condition $B_i \subseteq B_j$ is valid.

(2) Consider now the implication: $if \;\; B_i \subseteq B_j \;\;$ then $Per_{FS}(B_i \mapsto B_j) = 0$. Assume that

$$if \;\; B_i \subseteq B_j \;\; then \;\; \mu_{B_i}(v) \leq \mu_{B_j}(v) \;\; and \;\; \mu_{B_i \cap B_j}(v) = \mu_{B_i}(v), \;\; \forall v \in V.$$

By Eq. 6.28, we obtain $Per_{FS}(B_i \mapsto B_j) = 0$. Thereby, the equality $Per_{FS}(B_i \mapsto B_j) = 0$ is always verified $if \;\; B_i \subseteq B_j$.

Proof Item *(4)*

(1) Let us prove the inequality

$$0 \leq Per_{FS}(B_i \mapsto B_j) + Per_{FS}(B_j \mapsto B_i).$$

Due to Eq. 6.27, the sum of the perturbations

$$Per_{FS}(B_i \mapsto B_j) + Per_{FS}(B_j \mapsto B_i)$$

is non-negative.

(2) The right hand side of the inequality can be expressed as

$$Per_{FS}(B_i \mapsto B_j) + Per_{FS}(B_j \mapsto B_i) =$$

$$= \frac{\sum_{l=1}^{L}(\mu_{B_i}(v_l) - \mu_{B_i \cap B_j}(v_l))}{\sum_{l=1}^{L}(\mu_{B_i}(v_l) + \mu_{B_j}(v_l))} + \frac{\sum_{l=1}^{L}(\mu_{B_j}(v_l) - \mu_{B_j \cap B_i}(v_l))}{\sum_{l=1}^{L}(\mu_{B_j}(v_l) + \mu_{B_i}(v_l))} =$$

$$= \frac{\sum_{l=1}^{L}(\mu_{B_i}(v_l) - \mu_{B_i \cap B_j}(v_l) + \mu_{B_j}(v_l) - \mu_{B_j \cap B_i}(v_l))}{\sum_{l=1}^{L}(\mu_{B_i}(v_l) + \mu_{B_j}(v_l))}.$$

It can be noticed that the inequalities

$$\mu_{B_i}(v) - \mu_{B_i \cap B_j}(v) \leq \mu_{B_i}(v), \text{ and}$$

$$\mu_{B_j}(v) - \mu_{B_j \cap B_i}(v) \leq \mu_{B_j}(v),$$

are satisfied. The right hand side of the inequality can be expressed as

$$\frac{\sum_{l=1}^{L}(\mu_{B_i}(v_l) - \mu_{B_i \cap B_j}(v_l) + \mu_{B_j}(v_l) - \mu_{B_j \cap B_i}(v_l))}{\sum_{l=1}^{L}(\mu_{B_i}(v_l) + \mu_{B_j}(v_l))} \leq$$

$$\leq \frac{\sum_{l=1}^{L}(\mu_{B_i}(v_l) + \mu_{B_j}(v_l))}{\sum_{l=1}^{L}(\mu_{B_i}(v_l) + \mu_{B_j}(v_l))} = 1.$$

□

Again, let us consider two fuzzy sets $B_i, B_j \in FS(V)$ in the set V of nominal elements. We can prove other interesting properties of the notion of perturbation of one fuzzy set by another fuzzy set, presented here as Corollary 6.3.

Corollary 6.3 *The sum of the measures of perturbation of an arbitrary fuzzy set B_i and another fuzzy set B_j satisfies the following equality*

$$Per_{FS}(B_i \mapsto B_j) + Per_{FS}(B_j \mapsto B_i) = \frac{card(B_i \triangle B_j)}{card(B_i \oplus B_j}. \tag{6.30}$$

Proof The left hand side of the equation can be rewritten as follows

$$Per_{FS}(B_i \mapsto B_j) + Per_{FS}(B_j \mapsto B_i) = \frac{\sum_{l=1}^{L}(\mu_{B_i}(v_l) - \mu_{B_i \cap B_j}(v_l))}{\sum_{l=1}^{L}(\mu_{B_i}(v_l) + \mu_{B_j}(v_l))} +$$

$$+ \frac{\sum_{l=1}^{L}(\mu_{B_j}(v_l) - \mu_{B_j \cap B_i}(v_l))}{\sum_{l=1}^{L}(\mu_{B_i}(v_l) + \mu_{B_j}(v_l))} = \frac{\sum_{l=1}^{L} | \mu_{B_i}(v_l) - \mu_{B_j}(v_l) |}{\sum_{l=1}^{L}(\mu_{B_i}(v_l) + \mu_{B_j}(v_l))} = \frac{card(B_i \triangle B_j)}{card(B_i \oplus B_j}.$$

□

It seems to be interesting to notice that the right hand side of Eq. 6.30 determines a distance measure $D_{12}(B_i, B_j)$ (cf. [10, 12]),

$$Per_{FS}(B_i \mapsto B_j) + Per_{FS}(B_j \mapsto B_i) = D_{12}(B_i, B_j),$$

where we keep the original index notation from Sect. 6.2.3.

So, one can say that the sum of two coupled measures of perturbation of fuzzy sets, namely the measure of perturbation of the set B_j by the set B_i, denoted by $Per_{FS}(B_i \mapsto B_j)$, and the measure of perturbation of the fuzzy set B_i by the set

B_j, denoted by $Per_{FS}(B_j \mapsto B_i)$, is equivalent to a measure of dissimilarity of two fuzzy sets.

6.3.6 Graphical Illustration of Measures of Perturbations of Fuzzy Sets

Let us assume that $card(V) = 1$ (i.e., $V = \{v\}$), and we consider the following two fuzzy sets $A, B \in FS(V)$, denoted by

$$A = \{(v, \mu_A(v))\} \text{ and } B = \{(v, \mu_B(v))\}.$$

In order to demonstrate the differences between the measure of perturbation of a fuzzy set A by another fuzzy set B (i.e., $Per_{FS}(B \mapsto A)$), and the measure of perturbation of the fuzzy set B by the fuzzy set A (i.e., $Per_{FS}(A \mapsto B)$), we use some geometrical interpretation of the measure of perturbation of the fuzzy sets.

According to Eqs. 6.28 and 6.29, the measures of perturbation of the two fuzzy sets have the following forms:

$$Per_{FS}(A \mapsto B) = \frac{\mu_A(v) - \mu_{A\cap B}(v)}{\mu_A(v) + \mu_B(v)},$$

$$Per_{FS}(B \mapsto A) = \frac{\mu_B(v) - \mu_{B\cap A}(v)}{\mu_B(v) + \mu_A(v)}.$$

Figures 6.15, 6.16, 6.17 and 6.18 show the comparisons between the measures of perturbation $Per_{FS}(B \mapsto A)$ and $Per_{FS}(A \mapsto B)$, for two fuzzy sets A and B. The figures make it apparent how the measure changes with respect to the value $\mu_B(v)$ (the value of $\mu_B(v)$ is shifted from 0 to 1), while the value of the membership function $\mu_A(v)$ is fixed. Here, we present the description of these figures.

- Let us consider $Per_{FS}(B \mapsto A)$, for fixed $\mu_A(v) = 0.5$ and $\mu_B(v)$ changing from 0 to 1.

First, the measure of perturbation $Per_{FS}(B \mapsto A)$ for the fixed membership function

$$\mu_A(v) = 0.5,$$

and the values of the membership function $\mu_B(v)$ changing from 0 to 1, namely

$$\mu_B(v) = 0.0,\ 0.1,\ 0.2,\ 0.3,\ 0.4,\ 0.5,\ 0.6,\ 0.7,\ 0.8,\ 0.9,\ 1.0,$$

are shown in Fig. 6.15.

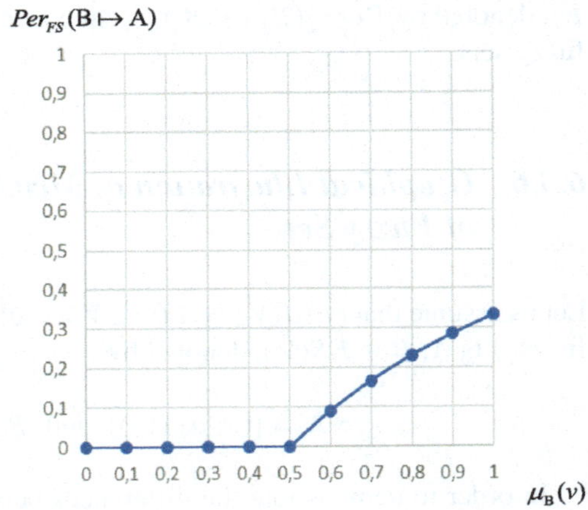

Fig. 6.15 The values of the measure of perturbation $Per_{FS}(B \mapsto A)$, for fixed value $\mu_A(v) = 0.5$ and $\mu_B(v)$ changes from 0 to 1

In Fig. 6.15, the measure of the perturbation is displayed, equal 0 for

$$\mu_A(v) = \mu_B(v) = 0.5$$

and for the changing membership function $\mu_B(v)$, where

$$0 \le \mu_B(v) \le \mu_A(v) = 0.5;$$

while for

$$0.5 < \mu_B(v) \le 1$$

the measure has the convex curvilinear form.

- Let us consider $Per_{FS}(B \mapsto A)$, for $\mu_A(v)$ changing from 0 to 1, and $\mu_B(v)$ changing from 0 to 1, with the step of 0.1.

The same graphical interpretation of the perturbation measure from Fig. 6.15 is repeated in Fig. 6.16, namely, the measure of perturbation $Per_{FS}(B \mapsto A)$ is drawn for different, fixed values of the membership function

$$\mu_A(v) = 0.0,\ 0.05,\ 0.1,\ 0.2,\ 0.3,\ 0.4,\ 0.5,\ 0.6,\ 0.7,\ 0.8,\ 0.9,\ 1.0$$

versus the similar series of values of the membership function $\mu_B(v)$, ranging from 0 to 1,

$$\mu_B(v) = 0.0,\ 0.05,\ 0.1,\ 0.2,\ 0.3,\ 0.4,\ 0.5,\ 0.6,\ 0.7,\ 0.8,\ 0.9,\ 1.0,$$

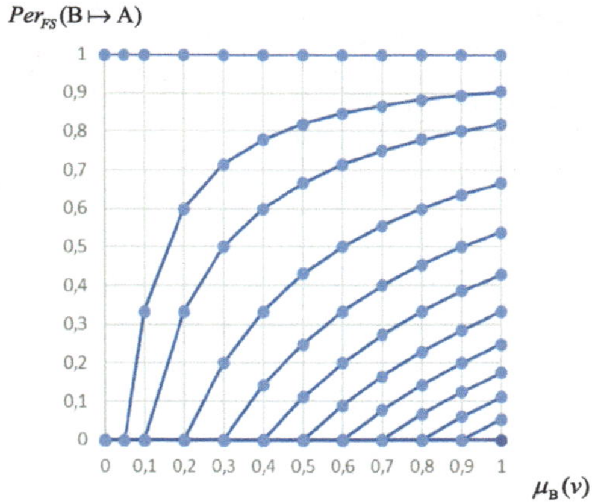

Fig. 6.16 The values of the measure of perturbation $Per_{FS}(B \mapsto A)$, where $\mu_A(v)$ changes from 0 to 1 and $\mu_B(v)$ changes from 0 to 1

see Fig. 6.16. A single plot is determined for one, fixed value of the membership function $\mu_A(v)$ (which changes for the next plot) and the changing values of the membership function $\mu_B(v)$, ranging from 0 to 1.

It is easy to notice in Fig. 6.16 that the shapes of plots of the perturbation measures $Per_{FS}(B \mapsto A)$ are very similar in their character, namely the first part of each plot is flat and equal 0 for

$$\mu_B(v) \leq \mu_A(v),$$

while the values of the perturbation measure increase for

$$\mu_B(v) > \mu_A(v).$$

Additionally, the value of the parameter $\mu_B(v)$ causes the shift of the plots in vertical direction.

- Let us consider $Per_{FS}(A \mapsto B)$ for fixed $\mu_A(v) = 0.5$ and $\mu_B(v)$ changing from 0 to 1, with the step of 0.1.

An analogous graphical interpretation of the opposite measure of perturbation $Per_{FS}(A \mapsto B)$ for fixed

$$\mu_A(v) = 0.5,$$

where the value of membership function $\mu_B(v)$ is changed from 0 to 1, namely

$$\mu_B(v) = 0.0,\ 0.1,\ 0.2,\ 0.3,\ 0.4,\ 0.5,\ 0.6,\ 0.7,\ 0.8,\ 0.9,\ 1.0,$$

is shown in Fig. 6.17.

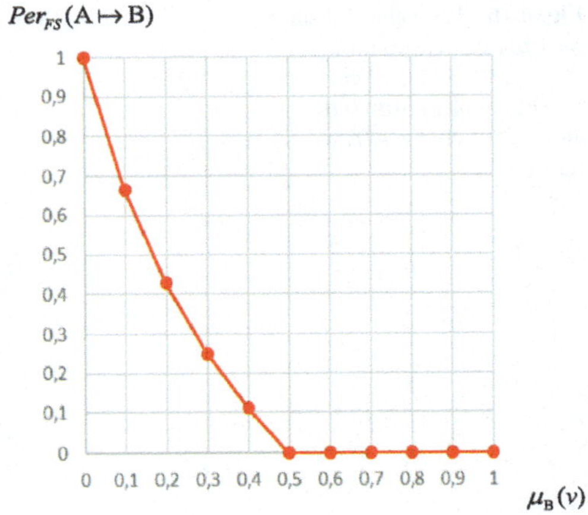

Fig. 6.17 The values of the measure of perturbation $Per_{FS}(A \mapsto B)$, where $\mu_A(v) = 0.5$ and $\mu_B(v)$ is changed from 0 to 1, with the step of 0.1

The first part of the plot (see Fig. 6.17) is concave, while the second is flat and equal 0.

- Let us consider $Per_{FS}(A \mapsto B)$, for $\mu_A(v)$ changing from 0 to 1, and $\mu_B(v)$ changing from 0 to 1, with the step of 0.1.

In the subsequent figure, the plot from Fig. 6.17 (for one fixed $\mu_A(v) = 0.5$) is drawn for different, fixed values of the membership function $\mu_A(v)$, namely

$$\mu_A(v) = 0.0,\ 0.1,\ 0.2,\ 0.3,\ 0.4,\ 0.5,\ 0.6,\ 0.7,\ 0.8,\ 0.9,\ 1.0$$

while the value of membership function $\mu_B(v)$ is changed from 0 to 1, as follows:

$$\mu_B(v) = 0.0,\ 0.1,\ 0.2,\ 0.3,\ 0.4,\ 0.5,\ 0.6,\ 0.7,\ 0.8,\ 0.9,\ 1.0.$$

In effect, the family of curves of the perturbation measure $Per_{FS}(A \mapsto B)$ for different values of the respective memberships functions is shown in Fig. 6.18.

Thus, when looking at the pair of figures, Figs. 6.15 and 6.17, one can observe totally different characteristics of fuzzy set perturbation measures $Per_{FS}(B \mapsto A)$ and $Per_{FS}(A \mapsto B)$. Plots in Figs. 6.16 and 6.18 just confirm this observation, and some properties of the perturbation measures will be studied in the forthcoming sections of this book.

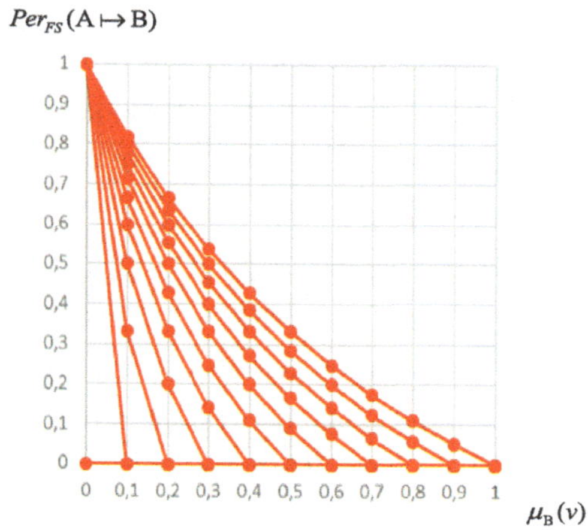

Fig. 6.18 The values of the measures of perturbation $Per_{FS}(A \mapsto B)$, where $\mu_A(v)$ is changed from 0 to 1, and $\mu_B(v)$ is changed from 0 to 1

6.4 Illustrative Examples

It seems to be interesting to compare the measures developed here with some selected distances for fuzzy sets considering the following examples.

6.4.1 Comparison of Selected Proximity Measures

The green sea turtles inhabit tropical and subtropical coastal waters around the world. They are listed as endangered (facing a very high risk of extinction). Despite this, they are still killed for their meat and eggs and their nesting grounds are disturbed by human encroachment. In the protected environment of a marine zoological park or farm, scientists can examine aspects of the sea turtles biology that are difficult or impossible to study in the wild, but the green turtle ranching and farming are still in the pioneer stages of development. Samples of a turtle's diet can be retrieved by stomach flushing. Green sea turtles change their diet during their ontogeny from omnivorous diet to herbivorous. This change in food preference is not uncommon, since it has been reported in other reptiles [26]. Green turtles are the only sea turtles that are herbivorous in adulthood. Young green sea turtles are mainly carnivorous and eat invertebrates such as sponges, jellyfish, aquatic insects, young crustaceans, crabs and worms. However, they gradually shift to an entirely vegetarian diet as they exit their juvenile stage. The jaws of green sea turtles are finely serrated, an adaptation fit for their diet of sea grasses and algae.

The problem is to calculate the degrees of proximity between the diets of the turtles, compare the diets in the study population and formulate some conclusions about the age of the turtles.

Let us describe the turtles in the form of fuzzy sets, corresponding to their diets. Let us consider the examples of the two turtles' diets as fuzzy sets, namely

A – "diet of the green turtle #1"
B – "diet of the green turtle #2"

in the following set

$$V = \{sea\ grass, algae, seaweed, crustaceans, aquatic\ insects, jellyfish, crabs, sponges, turtles\}.$$

The two fuzzy sets considered are described in detail as follows:

$$A = \{(sea\ grass, 0.9), (algae, 1.0), (seaweed, 0.8), (crustaceans, 0.0), (aquatic\ insects, 0.0), (jellyfish, 0.0), (crabs, 0.0), (sponges, 0.0), (turtles, 0.0)\},$$

$$B = \{(sea\ grass, 0.6), (algae, 0.8), (seaweed, 0.7), (crustaceans, 0.4), (aquatic\ insects, 0.5), (jellyfish, 0.3), (crabs, 0.2), (sponges, 0.4), (turtles, 0.0)\}.$$

In general, there does not exist a single "best" measure of proximity between two fuzzy sets. Many, proximity measures known from the literature, were developed specially for the considered data and the stated problems. Let us consider some selected measures of set similarity [8], namely

$$S_1(A, B),\ S_3(A, B),\ S_4(A, B),\ S_5(A, B),\ S_9(A, B),\ S_{12}(A, B),\ S_{14}(A, B),$$

where we keep the original notation from Sect. 6.2.3. Next, we assume that for two fuzzy sets A and B, the measures of dissimilarity is expressed by:

$$D_1(A, B) = 1 - S_1(A, B),$$
$$D_3(A, B) = 1 - S_3(A, B),$$
$$D_4(A, B) = 1 - S_4(A, B),$$
$$D_5(A, B) = 1 - S_5(A, B),$$
$$D_9(A, B) = 1 - S_9(A, B),$$
$$D_{12}(A, B) = 1 - S_{12}(A, B),$$
$$D_{14}(A, B) = 1 - S_{14}(A, B).$$

The graphic illustration of the calculated measures of dissimilarity and measures of perturbation $Per_{FS}(A \mapsto B)$ and $Per_{FS}(B \mapsto A)$ is provided in Fig. 6.19.

It is easy to notice, for the considered two fuzzy sets, that different measures of fuzzy set proximity generate different respective values, in general. Therefore, it seems that it is impossible to indicate which measure is better in general. It means that there does not exist the best measure for evaluation of proximity between two arbitrary fuzzy sets and the choice depends on the nature of data under consideration.

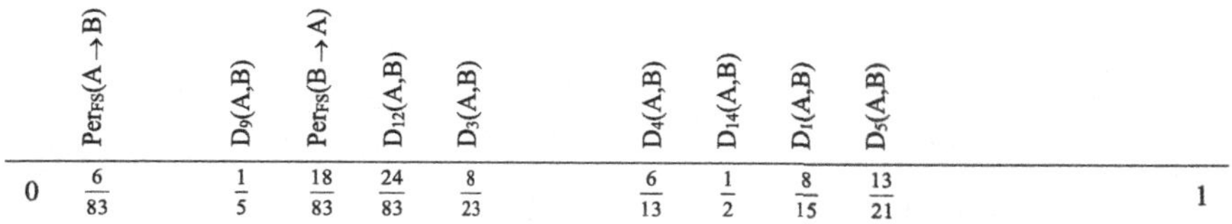

Fig. 6.19 Graphical illustration of selected measures

Yet, regarding this example, the proposed measure of fuzzy set perturbation allows us to compare the diets of turtles and we can formulate some additional conclusions concerning their age.

The measure of perturbation of the fuzzy set B by the fuzzy set A has a lower value than the measure of perturbation of the fuzzy set A by the fuzzy set B ($6/83$ vs. $18/83$). So, we can say that the diet of the green turtle #1 (described by the fuzzy set A) is less varied than the diet of the turtle #2 (described by the fuzzy set B). The diet of the turtle #2 is more diversified and assuredly this animal is younger than turtle #1, a conclusion, which could not be formulated based on symmetric measures.

Thereby, this short example highlights the differences between the selected proximity measures (which are obviously symmetric) and the fuzzy set perturbations (which are not necessarily symmetric).

6.4.2 Measures of Perturbations Versus Hamming Distance

It seems to be interesting to compare the measures of the fuzzy set perturbation and the Hamming distance for fuzzy sets. First, let us assume that $card(V) = 1$ (i.e., $V = \{v\}$), and consider the following two fuzzy sets $A, B \in FS(V)$, denoted by

$$A = \{(v, \mu_A(v))\} \text{ and } B = \{(v, \mu_B(v))\}.$$

The Hamming distance between two fuzzy sets is expressed as follows:

$$D_{nHamming}(A, B) = \mid \mu_A(v) - \mu_B(v) \mid .$$

According to Eqs. 6.28 and 6.29, the measures of perturbation of the two fuzzy sets have the following forms:

$$Per_{FS}(A \mapsto B) = \frac{\mu_A(v) - \mu_{A\cap B}(v)}{\mu_A(v) + \mu_B(v)},$$

and

$$Per_{FS}(B \mapsto A) = \frac{\mu_B(v) - \mu_{B\cap A}(v)}{\mu_B(v) + \mu_A(v)}.$$

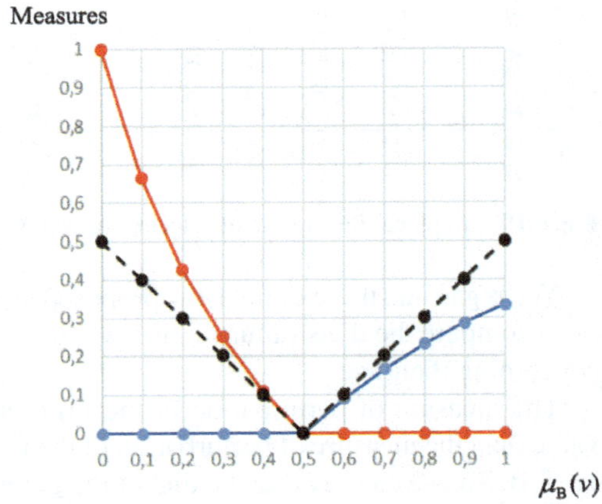

Fig. 6.20 Comparison between the values of Hamming distance $D_{nHamming}(A, B)$ (marked with dotted line) and $Per_{FS}(A \mapsto B)$, marked in red, as well as $Per_{FS}(B \mapsto A)$, marked in blue

Figure 6.20 shows the comparison between the measures of perturbation $Per_{FS}(A \mapsto B)$ and $Per_{FS}(B \mapsto A)$ and the Hamming distance $D_{nHamming}(A, B)$, for the following two exemplary fuzzy sets

$$A = \{(v, 0.5)\},$$

$$B = \{(v, \mu_B(v))\},$$

where $\mu_B(v)$ changes from 0 to 1, with the step 0.1. The plots are drawn with respect to the values of the membership function $\mu_B(v)$, assuming that the membership function $\mu_A(v)$ is fixed.

The measures of perturbation are just the same as shown in Figs. 6.15 and 6.17, meaning the measure of perturbation of the fuzzy set A by the fuzzy set B, $Per_{FS}(B \mapsto A)$ (depicted in Fig. 6.15), and the measure of perturbation of the fuzzy set B by the fuzzy set A, $Per_{FS}(A \mapsto B)$ (depicted in Fig. 6.17), and they are marked by solid lines in Fig. 6.20. The measures of perturbation $Per_{FS}(A \mapsto B)$ are marked in red, and the measures of perturbation $Per_{FS}(B \mapsto A)$ are marked in blue.

The values of the Hamming distance between the fuzzy set A and the fuzzy set B, denoted by $D_{nHamming}(A, B)$, are calculated and are marked by a dotted line in Fig. 6.20.

This short example highlights the differences between the Hamming distance (which is symmetric) and the measures of perturbation (which are not necessarily symmetrical).

References

1. J. Łukasiewicz, O logice trójwartościowej, Ruch filozoficzny, 1920, 5, 170–171, English translation On three-valued logic, in *Selected Works*. ed. by L. Borkowski (North-Holland, Amsterdam, 1970), pp.87–88
2. J. Łukasiewicz, *Elementy logiki matematycznej, PWN, Warszawa, 1929* (Pergamon Press, Oxford, 1963). English translation Elements of Mathematical Logic
3. L.A. Zadeh, Fuzzy sets. Inf. Control **8**, 338–353 (1967)
4. J. Kacprzyk, *Fuzzy Sets in System Analysis* (PWN, Warsaw, 1986)
5. J. Kacprzyk, *Multistage Fuzzy Control: A Model-Based Approach to Fuzzy Control and Decision Making* (Wiley, Inc, 1997)
6. J.M. Mendel, Type-2 fuzzy sets and systems: an overview. IEEE Comput. Intell. Mag. **2**, 20–29 (2007)
7. R. Moore, W. Lodwick, Interval analysis and fuzzy set theory. Fuzzy Sets Syst. **135**(1), 5–9 (2003)
8. L. Baccour, A.M. Alimi, R.I. John, Some notes on fuzzy similarity measures and application to classification of shapes, recognition of Arabic Sentences and Mosaic. IAENG Int. J. Comput. Sci. **41**(2), 81–90 (2014)
9. D. Dubois, H. Prade, *Fuzzy Sets and Systems, Theory and Applications* (Academic Press, New York, 1980)
10. C.P. Pappis, I. Karacapilidis, A comparative assessment of measures of similarity of fuzzy values. Fuzzy Sets Syst. **56**, 171–174 (1993)
11. D.V.D. Weken, M. Nachtegael, E.E. Kerre, Using similarity measures and homogeneity for the comparison of images. Image Vis. Comput. **22**, 695–702 (2004)
12. D.-G. Wang, Y.-P. Meng, H.-X. Li, A fuzzy similarity inference method for fuzzy reasoning. Comput. Math. Appl. **56**, 2445–2454 (2008)
13. V.V. Cross, T.A. Sudkamp, *Similarity and Compatibility in Fuzzy Set Theory* (Physica-Verlag, Heidelberg, 2002)
14. J. Kacprzyk, M. Krawczak, G. Szkatuła, On bilateral matching between fuzzy set. Inf. Sci. **402**, 244–266 (2017)
15. L.A. Zadeh, The concept of a linguistic variable and its application to approximate reasoning. Inf. Sci. **8**, 199–249 (1975)
16. B. Bouchon-Meunier, M. Rifqi, S. Bothorel, Towards general measures of comparison of objects. Fuzzy Sets Syst. **84**, 143–153 (1996)
17. P.Z. Wang, *Fuzzy Sets and its Applications* (Shanghai Science and Technology Press, Shanghai, 1983)
18. W.-J. Wang, New similarity measures on fuzzy sets and on elements. Fuzzy Sets Syst. **85**, 305–309 (1997)
19. X. Liu, Entropy, distance measure and similarity measure of fuzzy sets and their relations. Fuzzy Sets Syst. **52**(3), 305–318 (1992)
20. S. Raha, A. Hossain, S. Ghosh, Similarity based approximate reasoning: fuzzy control. J. Appl. Log. **6**, 47–71 (2008)
21. J. Fan, W. Xie, Some notes on similarity measures and proximity measures. Fuzzy Sets Syst. **101**, 403–412 (1999)
22. G.J. Klir, T.A. Folger, *Fuzzy Sets. Uncertainty and Information* (Prentice Hall, Englewood Cliffs, 1988)
23. M. Nowakowska, Methodological problem of measurement of fuzzy concept in the social sciences. Syst. Res. Behav. Sci. **22**(2), 107–115 (1977)
24. A. Tversky, Features of similarity. Psychol. Rev. **84**(4), 327–352 (1977)
25. D. Dubois, H. Prade, Fuzzy cardinality and the modeling of imprecise quantification. Fuzzy Sets Syst. **16**, 199–230 (1985)
26. H.F. Hirth, Synopsis of biological data on the green turtle. Biol. Rep. **97**(1), (Department of Biology University of Utah, 1971)

Chapter 7
Matching Between Intuitionistic Fuzzy Sets

7.1 Preliminaries

The essential extension to fuzzy sets is known as *intuitionistic fuzzy sets* (or *Atanassov intuitionistic fuzzy sets*, or IFS sets), and was developed by Krassimir Atanassov in 1983 [1]. Atanassov considered not only the degree of membership but also the degree of non-membership for any element belonging to such sets. If the sum of theses degrees is less than one, then the remainder (according to Atanassov) is called the degree of indeterminacy of information (or hesitation, or Atanassov index).

There is a natural extension of intuitionistic fuzzy sets, called the interval-valued intuitionistic fuzzy sets ($IVIFS$), introduced by Atanassov [2]. In $IVIFS$s, the certainty as well as the uncertainty, and the indeterminacy are represented by interval values.

The use of fuzzy sets, or interval-valued fuzzy sets, or intuitionistic fuzzy sets, or interval-valued intuitionistic fuzzy sets depends to an essential degree on the considered problem.

In this book, we consider a non-empty, finite set of objects, concepts or patterns described by a set of nominal-valued attributes, and in this chapter, the values of the attributes are not precisely known. Additionally, in this chapter it is assumed that there is some information, whose value is possible or acceptable, and some information, which is characterized as impossible or unacceptable.

It is interesting that application of the intuitionistic fuzzy sets theory allows us to model such imperfect, incomplete and inconsistent data. Additionally, in the case of such sets there is also some degree of information indeterminacy. Thereby, the intuitionistic fuzzy sets may have potential applications in modeling with respect to many different fields, like artificial intelligence, computer science, economics, quantum computing, medical science, etc.

In order to develop effective procedures for comparing objects described by intutionistic fuzzy sets, we consider a new measure of proximity between two intuitionis-

G. Szkatuła and M. Krawczak, *Bidirectional Comparison of Nominal Sets*,
Studies in Computational Intelligence 1140,
https://doi.org/10.1007/978-3-031-53096-8_7

tic fuzzy sets. Namely, we refer to the so called perturbation of one intuitionistic fuzzy set by another intuitionistic fuzzy set as well as a measure of such perturbation. Additionally, we can verify the symmetry of such objects' proximity. Our considerations are based on the intuitionistic fuzzy sets theory and its basic operations. We consider commonly used similarity or dissimilarity measures between two intuitionistic fuzzy sets and compare it with the perturbation measure. After normalization, the value of the measure of perturbation of intuitionistic fuzzy sets ranges from 0 to 1, where 0 is the lowest value of perturbation while 1 is the highest value of perturbation. It is worth to notice that this newly developed measure is not always symmetrical. The asymmetry is meant as follows: the value of the measure of perturbation of the first intuitionistic fuzzy set by the second intuitionistic fuzzy set can be different from the value of the measure of perturbation of the second intuitionistic fuzzy set by the first intuitionistic fuzzy set. In this way, the two measures of perturbation cannot be considered as the distance between the intuitionistic fuzzy sets. Additionally, the sum of these measures gives an equivalent interpretation of dissimilarity measure between two intuitionistic fuzzy sets.

Once again we would like to mention that the term *impact* or *perturbation* is used in the general sense (i.e., as a disturbance of one set by another set). A good explanation is contained in the paper by Kacprzyk et al. [3]. Thus, the term of impact of one set on another set or of perturbation of one set by another set corresponds to Tversky's considerations about object similarities.

The concept of perturbation of the intuitionistic fuzzy sets can find wide applications in solving problems, related to comparison of such sets, when the direction of comparison of the sets is significant. Thus, the concept of the measure of set perturbation is obviously more general than the usual proximity or distance measures.

7.2 Brief Introduction to the Intuitionistic Fuzzy Sets Theory

We would like to recall now some basic notions of one of the extensions of the fuzzy sets theory, namely the intuitionistic fuzzy sets theory (see e.g., Atanassov [4, 5]; Atanasow and Gargov [2]).

Again, let V be any finite and non-empty set.

Definition 7.1 (*Intuitionistic fuzzy set, 2D representation*) Let us consider a non-empty and finite set V of nominal, distinct elements, $V = \{v_1, v_2, \ldots, v_L\}$. The intuitionistic fuzzy set C in the crisp set V (in the sense of Atanassov) is defined as the set of ordered triples of the form

$$C = \{(v, \mu_C(v), \nu_C(v)) \mid \forall v \in V\} \tag{7.1}$$

where $\mu_C : V \to [0, 1]$ is the membership function of the intuitionistic fuzzy set C, and $\nu_C : V \to [0, 1]$ is the non-membership function, with an additional condition

$$0 \le \mu_C(v) + \nu_C(v) \le 1$$

for each $v \in V$.

The numbers $\mu_C \in [0, 1]$ and $\nu_C \in [0, 1]$ determine, first, *the degree of membership* of the element $v \in V$ to the intuitionistic fuzzy set C and, second, *the degree of non − membership* of the element $v \in V$ to the intuitionistic fuzzy set C, respectively. The collection of all intuitionistic fuzzy sets in the crisp set V of nominal elements is denoted by $IFS(V)$.

The value

$$\pi_C(v) = 1 - \mu_C(v) - \nu_C(v)$$

is called *the degree of indeterminacy* of information (or *hesitation*, or *Atanassov's index*).

According to Eq. 7.1, for example, the set

$$C = \{(v_1,\ 0.3,\ 0.5), (v_2,\ 0.6,\ 0.1), (v_3,\ 0.5,\ 0.3)\}$$

can be considered as an intuitionistic fuzzy set in $V = \{v_1, v_2, v_3\}$.

In this way, each element of the intuitionistic fuzzy set (7.1) can be represented as a point having two coordinates, as follows

$$\left(\mu_C(v),\ \nu_C(v)\right).$$

The above two-dimensional geometrical interpretation of the elements of an intuitionistic fuzzy set is shown in Fig. 7.1.

The point $\left(\mu_C(v), \nu_C(v)\right) = (0, 0)$ represents the element $v \in V$, for which the degree of indeterminacy equals 1, and nothing can be said about belonging or not belonging to the corresponding intuitionistic fuzzy set C. Another characteristic point $(1, 0)$ represents the element v, which belongs for sure to the intuitionistic fuzzy set C. Point $(0, 1)$ represents the element v, which for sure does not belong to the fuzzy set C.

Any other combination of the membership degree and a non-membership degree represents an element v, subject to the following constraints

$$0 \le \mu_C(v) + \nu_C(v) \le 1,$$

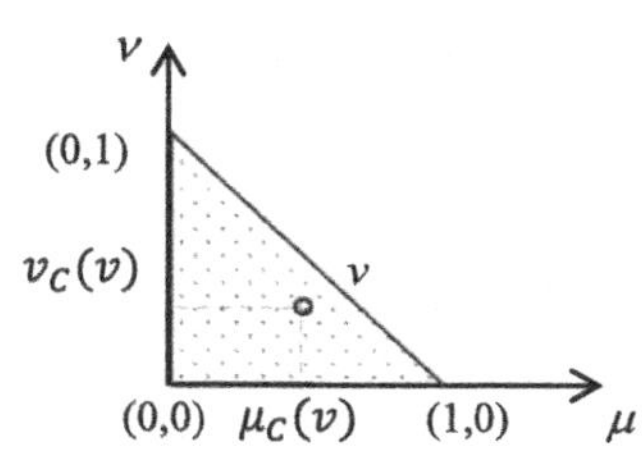

Fig. 7.1 Geometrical interpretation of an element of an intuitionistic fuzzy set (in two-dimensional space), $v \in V$ [1]

this element belonging to some considered intuitionistic fuzzy set C. Thus, all points v inside the triangle belong to the intuitionistic fuzzy set C.

In the case when $\forall v \in V$

$$\nu_C(v) = 1 - \mu_C(v),$$

then such an intuitionistic fuzzy set C is reduced to a fuzzy set, characterized by the membership function $\mu_C(v)$.

Another approach to the intuitionistic fuzzy set takes into account explicitly the degree of hesitation margin. This model was introduced by Szmidt and Kacprzyk [6].

Definition 7.2 (*Intuitionistic fuzzy set, 3D representation*) Let us consider a non-empty and finite set V of nominal, distinct elements, $V = \{v_1, v_2, ..., v_L\}$. An intuitionistic fuzzy set C in the crisp set V (in the sense of Atanassov) is defined as the set of ordered fours of the form

$$C = \{(v,\ \mu_C(v),\ \nu_C(v),\ \pi_C(v)) \mid \forall v \in V\} \tag{7.2}$$

where $\mu_C : V \to [0, 1]$ is the membership function of the intuitionistic fuzzy set C, $\nu_C : V \to [0, 1]$ is the non-membership function, and $\pi_C : V \to [0, 1]$ denotes the degree of information indeterminacy, with an additional condition

$$0 \leq \mu_C(v) + \nu_C(v) \leq 1,$$

where $\pi_C(v) = 1 - \mu_C(v) - \nu_C(v)$, for each $v \in V$.

Thus, each element of the intuitionistic fuzzy set (7.2) can be represented as a point having three coordinates, as follows

$$(\mu_C(v),\ \nu_C(v),\ \pi_C(v)).$$

The three-dimensional geometrical interpretation is shown in Fig. 7.2. Due to the condition

$$\mu_C(v) + \nu_C(v) + \pi_C(v) = 1,$$

for any $v \in V$, each point inside the triangle can be treated as a representation of a certain element v belonging to the intuitionistic fuzzy set C. Sometimes, for simplicity, the value $v \in V$ can be omitted, when it does not lead to confusion in this book, and so, any point here considered has three coordinates

$$(\mu_C, \nu_C, \pi_C),$$

as shown in Fig. 7.2.

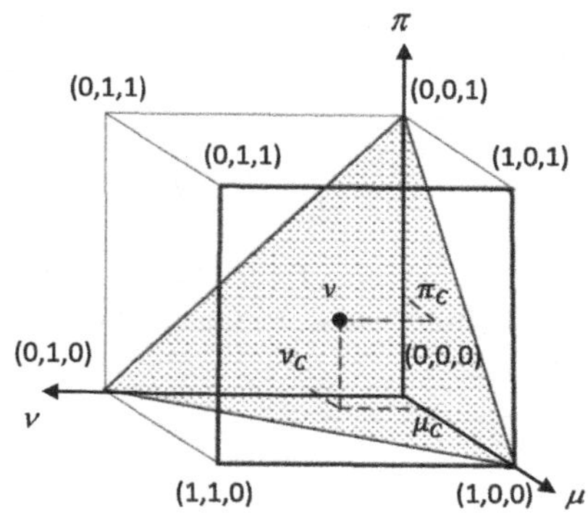

Fig. 7.2 Geometrical interpretation of an intuitionistic fuzzy set (in three-dimensional space), $v \in V$

In Fig. 7.2 there are some characteristic points, namely: point (1, 0, 0) represents the element v which belongs for sure to the intuitionistic fuzzy set C; the point (0, 1, 0) represents the element v which for sure does not belong to the intuitionistic fuzzy set; while the point (0, 0, 1) represents the element v which is absolutely unspecified as to whether it belongs or not to the intuitionistic fuzzy set C.

7.2.1 Operations on Intuitionistic Fuzzy Sets

Let us consider two intuitionistic fuzzy sets, C_i and C_j, in a crisp set $V = \{v_1, v_2, \ldots, v_L\}$, noted as follows:

$$C_i = \{(v,\ \mu_{C_i}(v),\ \nu_{C_i}(v)) \mid v \in V\},$$

$$C_j = \{(v,\ \mu_{C_j}(v),\ \nu_{C_j}(v)) \mid v \in V\}.$$

The union, intersection and complement of two intuitionistic fuzzy sets and other basic operations are given [4].

- *the union of the intuitionistic fuzzy sets* C_i and C_j is an intuitionistic fuzzy set, with the degree of membership $\mu_{C_i \cup C_j}(v)$, and non-membership $\nu_{C_i \cup C_j}(v)$ of the elements $v \in V$ of the set $C_i \cup C_j$, specified as

$$\left(\mu_{C_i \cup C_j}(v),\ \nu_{C_i \cup C_j}(v)\right) := (max\{\mu_{C_i}(v), \mu_{C_j}(v)\},\ min\{\nu_{C_i}(v), \nu_{C_j}(v)\});$$

- *the intersection of the intuitionistic fuzzy sets* C_i and C_j is an intuitionistic fuzzy set $C_i \cap C_j$, defined as

$$\left(\mu_{C_i \cap C_j}(v),\ \nu_{C_i \cap C_j}(v)\right) := (min\{\mu_{C_i}(v), \mu_{C_j}(v)\},\ max\{\nu_{C_i}(v), \nu_{C_j}(v)\});$$

- *the complement of an intuitionistic fuzzy set* C_i (denoted by $\neg C_i$ or C_i^C), is an intuitionistic fuzzy set

$$\left(\mu_{C_i^c}(v),\ \nu_{C_i^c}(v)\right) := (\nu_{C_i}(v), \mu_{C_i}(v)\});$$

- *the addition of the intuitionistic fuzzy sets* C_i and C_j, leads to an intuitionistic fuzzy set $C_i \oplus C_j$, defined as

$$\left(\mu_{C_i \oplus C_j}(v),\ \nu_{C_i \oplus C_j}(v)\right) :=$$

$$= (\mu_{C_i}(v) + \mu_{C_j}(v) - \mu_{C_i}(v) \cdot \mu_{C_j}(v),\ \ \nu_{C_i}(v) \cdot \nu_{C_j}(v));$$

- *the multiplication of the intuitionistic fuzzy sets* C_i and C_j, leads to an intuitionistic fuzzy set $C_i \otimes C_j$, defined as

$$\left(\mu_{C_i \otimes C_j}(v),\ \nu_{C_i \otimes C_j}(v)\right) :=$$

$$= \left(\mu_{C_i}(v) \cdot \mu_{C_j}(v),\ \ \nu_{C_i}(v) + \nu_{C_j}(v) - \nu_{C_i}(v) \cdot \nu_{C_j}(v)\right);$$

- *the symmetric difference of the intuitionistic fuzzy sets* C_i and C_j leads to an intuitionistic fuzzy set $C_i \bigtriangleup C_j$, defined as

$$\left(\mu_{C_i \bigtriangleup C_j}(v),\ \nu_{C_i \bigtriangleup C_j}(v)\right) :=$$

$$= \left(max\left\{min\{\mu_{C_i}(v), \nu_{C_j}(v)\},\ min\{\mu_{C_j}(v), \nu_{C_i}(v)\}\right\},\right.$$

$$\left. min\left\{max\{\nu_{C_j}(v), \mu_{C_i}(v)\},\ max\{\nu_{C_i}(v), \mu_{C_j}(v)\}\right\}\right).$$

7.2.2 Examples of Proximity Measures of Intuitionistic Fuzzy Sets

The literature reports many measures of proximity between the intuitionistic fuzzy sets $C_i, C_j \in IFS(V)$, and suitable formulas to calculate them. If only the degrees of membership $\mu_{C_i}(v), \mu_{C_j}(v)$ and non-membership $\nu_{C_i}(v),\ \nu_{C_j}(v)$ are considered, then a two-dimensional (2D) distance is used (e.g., Atanassov [5]). If the hesitancy degrees $\pi_{C_i}(v)$ and $\pi_{C_j}(v)$ are additionally taken into account, then a three-dimensional (3D) distance is used (e.g., Szmidt and Kacprzyk [6]).

Two − dimensional representation

Let us assume that a measure of distance is defined by a mapping

$$D : IFS(V) \times IFS(V) \to R^{+} \cup \{0\}.$$

We present below the most popular definitions using two-dimensional interpretation [5].

- *Hamming distance* between the intuitionistic fuzzy sets C_i and C_j is defined by

$$D_{Hamming}(C_i, C_j) = \frac{1}{2}\sum_{l=1}^{L}(\mid \mu_{C_i}(v_l) - \mu_{C_j}(v_l) \mid + \mid \nu_{C_i}(v_l) - \nu_{C_j}(v_l) \mid).$$

Note that condition $D_{Hamming}(C_i, C_j) \leq L$ is satisfied.

- *Euclidean distance* between the intuitionistic fuzzy sets C_i and C_j is defined by

$$D_{Euclidean}(C_i, C_j) = \sqrt{\frac{1}{2}\sum_{l=1}^{L}((\mu_{C_i}(v_l) - \mu_{C_j}(v_l))^2 + (\nu_{C_i}(v_l) - \nu_{C_j}(v_l))^2)}.$$

Note that condition $D_{Euclidean}(C_i, C_j) \leq \sqrt{L}$ is satisfied.

In turn let us assume, that a measure of distance is defined by a mapping

$$D : IFS(V) \times IFS(V) \to [0, 1].$$

We present below the most popular definitions.

- *The normalized Hamming distance* between the intuitionistic fuzzy sets C_i and C_j is defined by

$$D_{nHamming}(C_i, C_j) = \frac{D_{Hamming}(C_i, C_j)}{L} =$$

$$= \frac{1}{2L}\sum_{l=1}^{L}(\mid \mu_{C_i}(v_l) - \mu_{C_j}(v_l) \mid + \mid \nu_{C_i}(v_l) - \nu_{C_j}(v_l) \mid).$$

Note that condition $D_{nHamming}(C_i, C_j) \leq 1$ is satisfied.

- *The normalized Euclidean distance* between the intuitionistic fuzzy sets C_i and C_j is defined by

$$D_{nEuclidean}(C_i, C_j) = \frac{D_{Euclidean}(C_i, C_j)}{\sqrt{L}} =$$

$$= \sqrt{\frac{1}{2L}\sum_{l=1}^{L}((\mu_{C_i}(v_l) - \mu_{C_j}(v_l))^2 + (\nu_{C_i}(v_l) - \nu_{C_j}(v_l))^2)}.$$

Note that condition $D_{nEuclidean}(C_i, C_j) \leq 1$ is satisfied.

Three-dimensional representation

Based on three-dimensional representation, Szmidt and Kacprzyk [6] modified Atanassov's distances by including the third parameters, $\pi_{C_i}(v)$ and $\pi_{C_j}(v)$. We present below the examples of distances using three-dimensional interpretation.

- *Hamming distance* between the intuitionistic fuzzy sets C_i and C_j is defined by

$$D_{Hamming}(C_i, C_j) =$$

$$\frac{1}{2}\sum_{l=1}^{L}(\mid \mu_{C_i}(v_l) - \mu_{C_j}(v_l) \mid + \mid \nu_{C_i}(v_l) - \nu_{C_j}(v_l n) \mid + \mid \pi_{C_i}(v_l) - \pi_{C_j}(v_l) \mid).$$

  Note that condition $D_{Hamming}(C_i, C_j) \leq L$ is satisfied.

- *The normalized Hamming distance* between the intuitionistic fuzzy sets C_i and C_j is defined by

$$D_{nHamming}(C_i, C_j) =$$

$$\frac{1}{2L}\sum_{l=1}^{L}(\mid \mu_{C_i}(v_l) - \mu_{C_j}(v_l) \mid + \mid \nu_{C_i}(v_l) - \nu_{C_j}(v_l) \mid + \mid \pi_{C_i}(v_l) - \pi_{C_j}(v_l) \mid).$$

  Note that condition $D_{nHamming}(C_i, C_j) \leq 1$ is satisfied.

- *Euclidean distance* between the intuitionistic fuzzy sets C_i and C_j is defined by

$$D_{Euclidean}(C_i, C_j) =$$

$$\sqrt{\frac{1}{2}\sum_{l=1}^{L}((\mu_{C_i}(v_l) - \mu_{C_j}(v_l))^2 + (\nu_{C_i}(v_l) - \nu_{C_j}(v_l))^2 + (\pi_{C_i}(v_l) - \pi_{C_j}(v_l))^2)}.$$

- *The normalized Euclidean distance* between the intuitionistic fuzzy sets C_i and C_j is defined by

$$D_{nEuclidean}(C_i, C_j) =$$

$$\sqrt{\frac{1}{2L}\sum_{l=1}^{L}((\mu_{C_i}(v_l) - \mu_{C_j}(v_l))^2 + (\nu_{C_i}(v_l) - \nu_{C_j}(v_l))^2 + (\pi_{C_i}(v_l) - \pi_{C_j}(v_l))^2)}.$$

7.2.3 Extensions of Fuzzy Sets

Given the different representations of fuzzy sets, we wish to present now the following simple illustrative example.

The problem is as follows: how to describe the diets of the turtles and how to compare them. For more information about the diets of the turtles - see Sect. 6.4.1.

Let us assume that we describe the turtles in the form of fuzzy sets, corresponding their diets. Thus, let us consider an exemplary fuzzy set C, for instance,

C - "diet of the green turtle #1"

describing the diet of $turtle$#1 in terms of a crisp set V, for example, the set

$$V = \{sea\ grasses,\ jellyfish,\ turtle\},$$

as a fuzzy set C, as follows

$$C = \{(sea\ grasses, 1), (jellyfish, 0.4), (turtle, 0)\}.$$

Below, we present several examples of different extensions of the here considered fuzzy set, namely:

- in the sense of theory of interval-valued fuzzy sets, say

$$C = \{(sea\ grasses, [1, 1]), (jellyfish, [0.4, 0.5]), (turtle, [0, 0])\};$$

- in the sense of theory of intuitionistic fuzzy sets, say

$$C = \{(sea\ grasses, 1, 0), (jellyfish, 0.4, 0.5), (turtle, 0, 1)\};\ \text{or}$$

$$C = \{(sea\ grasses, 1, 0, 0), (jellyfish, 0.4, 0.5, 0.1), (turtle, 0, 1, 0)\};$$

- in the sense of theory of interval-valued intuitionistic fuzzy sets, say

$$C = \{(sea\ grasses, [1, 1], [0, 0]), (jellyfish, [0.4, 0.45], [0.5, 0.55]),$$
$$(turtle, [0, 0], [1, 1])\}.$$

Thus, for the here considered fuzzy set C, different theories generate different representations. Therefore, the conclusion is straightforward, it seems that it is impossible to indicate unambiguously which representation is better in general terms. This means that there does not exist the best representation and the choice depends on the nature of data under consideration and the nature of the problem under consideration.

It is obvious that using different representations we must also use operations and specific features of that representation, and the results must always be compared with those obtained without using fuzzy sets those obtained using different sets of assumptions.

7.3 Fundamentals of Perturbation of Intuitionistic Fuzzy Sets

Let us assume that there is a collection of intuitionistic fuzzy sets $IFS(V)$ in the non-empty set V, where V is a finite crisp set of nominal, distinct elements, that is, $V = \{v_1, v_2, ..., v_L\}$.

Let us consider two intuitionistic fuzzy sets $C_i, C_j \in IFS(V)$, described as follows:

$$C_i = \{(v,\ \mu_{C_i}(v),\ \nu_{C_i}(v)) \mid \forall v \in V\},$$

$$C_j = \{(v,\ \mu_{C_j}(v),\ \nu_{C_j}(v)) \mid \forall v \in V\},$$

where $\mu_{C_i}, \mu_{C_j} : V \to [0, 1]$ and $\nu_{C_i}, \nu_{C_j} : V \to [0, 1]$, with additional conditions

$$0 \le \mu_{C_i}(v) + \nu_{C_i}(v) \le 1,$$

$$0 \le \mu_{C_j}(v) + \nu_{C_j}(v) \le 1,$$

for each value $v \in V$.

7.3.1 The Concept of Perturbation of Intuitionistic Fuzzy Sets

We consider here the concept of *perturbation of one intuitionistic fuzzy set by another intuitionistic fuzzy set* (cf. Krawczak and Szkatuła [7]). The concept of the intuitionistic fuzzy set perturbations coincides with the first operation of subtraction of two such sets (cf. Atanassov [1]).

Definition 7.3 (*Perturbation of intuitionistic fuzzy set*) Let us assume that $C_i, C_j \in IFS(V)$, $V = \{v_1, v_2, ..., v_L\}$. The perturbation of the intuitionistic fuzzy set C_j by the intuitionistic fuzzy set C_i, denoted by $(C_i \mapsto C_j)$, is defined as a new intuitionistic fuzzy set, in the following manner:

$$(C_i \mapsto C_j) = \{\big(v,\ \mu_{C_i \mapsto C_j}(v),\ \nu_{C_i \mapsto C_j}(v)\big) \mid v \in V\}, \tag{7.3}$$

where

$$\mu_{C_i \mapsto C_j}(v) := min\{\mu_{C_i}(v),\ \nu_{C_j}(v)\},$$

$$\nu_{C_i \mapsto C_j}(v) := max\{\nu_{C_i}(v),\ \mu_{C_j}(v)\}.$$

Definitely, (7.3) can be rewritten as follows

$$(C_i \mapsto C_j) = \{\big(v,\ min\{\mu_{C_i}(v),\ \nu_{C_j}(v)\},\ max\{\nu_{C_i}(v),\ \mu_{C_j}(v)\}\big) \mid v \in V\}.$$

For the opposite case, the perturbation of the intuitionistic fuzzy set C_i by the intuitionistic fuzzy set C_j is analogously defined as follows:

$$(C_j \mapsto C_i) = \{\big(v,\ \mu_{C_j \mapsto C_i}(v),\ \nu_{C_j \mapsto C_i}(v)\big) \mid v \in V\}, \tag{7.4}$$

where

$$\mu_{C_j \mapsto C_i}(v) := min\{\mu_{C_j}(v),\ \nu_{C_i}(v)\},$$

$$\nu_{C_j \mapsto C_i}(v) := max\{\nu_{C_j}(v),\ \mu_{C_i}(v)\}.$$

So, (7.4) can be rewritten as follows

$$(C_j \mapsto C_i) = \{(v,\ min\{\mu_{C_j}(v),\ \nu_{C_i}(v)\},\ max\{\nu_{C_j}(v),\ \mu_{C_i}(v)\}) \mid v \in V\}.$$

Now, we present three important, even if straightforward, properties of the perturbation of one intuitionistic fuzzy set by another intuitionistic fuzzy set. These properties are stated in the form of a corollary, given in the sequel.

The first property confirms that the perturbation of intuitionistic fuzzy set is a new intuitionistic fuzzy set.

The second property is related to the conditions of equality of the perturbation of two intuitionistic fuzzy sets.

When, the third property gives the relation between the normalized Hamming distance (see, e.g., [6]) of two intuitionistic fuzzy sets and the normalized Hamming distance of intuitionistic fuzzy sets being the perturbations of these two $IFSs$, obtained by using (7.3) and (7.4).

Corollary 7.22 *Let us consider two intuitionistic fuzzy sets* $C_i, C_j \in IFS(V)$, *and the perturbations* $(C_i \mapsto C_j), (C_j \mapsto C_i) \in IFS(V)$. *The following properties of the perturbation hold:*

(1) $0 \le \mu_{C_i \mapsto C_j}(v) + \nu_{C_i \mapsto C_j}(v) \le 1$;

(2) $(C_i \mapsto C_j) = (C_j \mapsto C_i)$ *if and only if* $\left(\mu_{C_j}(v) = \mu_{C_i}(v) \text{ and } \nu_{C_j}(v) = \nu_{C_i}(v) \text{ or } \mu_{C_j}(v) = \nu_{C_j}(v) \text{ and } \mu_{C_i}(v) = \nu_{C_i}(v)\right)$;

(3) *if* $\left(\mu_{C_i}(v) \le \nu_{C_j}(v) \text{ and } \nu_{C_i}(v) \ge \mu_{C_j}(v)\right)$, *then* $D_{nHamming}(C_i, C_j) = D_{nHamming}\left((C_i \mapsto C_j), (C_j \mapsto C_i)\right)$.

Proof Property *(1)*

It can be noticed that by definition 7.3 (i.e., Eq. 7.3), the left hand side of the inequality, $0 \le \mu_{C_i \mapsto C_j}(v) + \nu_{C_i \mapsto C_j}(v)$, is satisfied.

Let us prove the right hand side of inequality, $\mu_{C_i \mapsto C_j}(v) + \nu_{C_i \mapsto C_j}(v) \le 1$. Consider now four cases:

Case 1: Conditions $\mu_{C_i}(v) \le \nu_{C_j}(v)$ and $\nu_{C_i}(v) \le \mu_{C_j}(v)$ are satisfied. It should be noticed that inequality

$$\mu_{C_i \mapsto C_j}(v) + \nu_{C_i \mapsto C_j}(v) = \mu_{C_i}(v) + \mu_{C_j}(v) \le \nu_{C_j}(v) + \mu_{C_j}(v) \le 1$$

is satisfied.

Case 2: Conditions $\mu_{C_i}(v) \le \nu_{C_j}(v)$ and $\nu_{C_i}(v) \ge \mu_{C_j}(v)$ are satisfied. Note that we can obtain the following inequality

$$\mu_{C_i \mapsto C_j}(v) + \nu_{C_i \mapsto C_j}(v) = \mu_{C_i}(v) + \nu_{C_i}(v) \le 1.$$

Case 3: Conditions $\mu_{C_i}(v) \ge \nu_{C_j}(v)$ and $\nu_{C_i}(v) \le \mu_{C_j}(v)$ are satisfied. Note that we can obtain the following inequality

$$\mu_{C_i \mapsto C_j}(v) + \nu_{C_i \mapsto C_j}(v) = \nu_{C_j}(v) + \mu_{C_j}(v) \le 1.$$

Case 4: Conditions $\mu_{C_i}(v) \ge \nu_{C_j}(v)$ and $\nu_{C_i}(v) \ge \mu_{C_j}(v)$ are satisfied. Note that we can obtain the following inequality

$$\mu_{C_i \mapsto C_j}(v) + \nu_{C_i \mapsto C_j}(v) = \nu_{C_j}(v) + \nu_{C_i}(v) \le \mu_{C_i}(v) + \nu_{C_i}(v) \le 1.$$

Proof Property *(2)*

Let us consider two impacts (perturbations) of one intuitionistic fuzzy set on (from) another intuitionistic fuzzy set

$$(C_i \mapsto C_j) = \{(v,\ min\{\mu_{C_i}(v),\ \nu_{C_j}(v)\},\ max\{\nu_{C_i}(v),\ \mu_{C_j}(v)\}) \mid v \in V\},$$

$$(C_j \mapsto C_i) = \{(v,\ min\{\mu_{C_j}(v),\ \nu_{C_i}(v)\},\ max\{\nu_{C_j}(v),\ \mu_{C_i}(v)\}) \mid v \in V\}.$$

The first implication,

$$if\ (C_i \mapsto C_j) = (C_j \mapsto C_i)\ then\ \left(\mu_{C_j}(v) = \mu_{C_i}(v)\ and\ \nu_{C_j}(v) = \nu_{C_i}(v)\right)$$

is obvious by Definition 7.3 (i.e., Eq. 7.3).

Let us consider the second implication,

$$if\ (C_i \mapsto C_j) = (C_j \mapsto C_i)\ then\ \left(\mu_{C_j}(v) = \nu_{C_j}(v)\ and\ \mu_{C_i}(v) = \nu_{C_i}(v)\right).$$

Consider now four possible cases:

Case 1: Conditions $\mu_{C_i}(v) \leq \nu_{C_j}(v)$ and $\nu_{C_i}(v) \leq \mu_{C_j}(v)$ are satisfied. It can be noticed, from Eqs. 7.3 and 7.4, that

$$if\ (C_i \mapsto C_j) = (C_j \mapsto C_i)\ then\ \left(\mu_{C_i}(v) = \nu_{C_i}(v)\ and\ \mu_{C_j}(v) = \nu_{C_j}(v)\right).$$

Case 2: Conditions $\mu_{C_i}(v) \leq \nu_{C_j}(v)$ and $\nu_{C_i}(v) \geq \mu_{C_j}(v)$ are satisfied. It can be noticed, from Eqs. 7.3 and 7.4, that

$$if\ (C_i \mapsto C_j) = (C_j \mapsto C_i)\ then\ \left(\mu_{C_j}(v) = \mu_{C_i}(v)\ and\ \nu_{C_j}(v) = \nu_{C_i}(v)\right).$$

Case 3: Conditions $\mu_{C_i}(v) \geq \nu_{C_j}(v)$ and $\nu_{C_i}(v) \leq \mu_{C_j}(v)$ are satisfied. It can be noticed, from Eqs. 7.3 and 7.4, that

$$if\ (C_i \mapsto C_j) = (C_j \mapsto C_i)\ then\ \left(\nu_{C_j}(v) = \nu_{C_i}(v)\ and\ \mu_{C_j}(v) = \mu_{C_i}(v)\right).$$

Case 4: Conditions $\mu_{C_i}(v) \geq \nu_{C_j}(v)$ and $\nu_{C_i}(v) \geq \mu_{C_j}(v)$ are satisfied. It can be noticed, from Eqs. 7.3 and 7.4, that

$$if\ (C_i \mapsto C_j) = (C_j \mapsto C_i)\ then\ \left(\mu_{C_j}(v) = \nu_{C_j}(v)\ and\ \mu_{C_i}(v) = \nu_{C_i}(v)\right).$$

In the opposite direction, the first implication states

$$if\ \left(\mu_{C_j}(v) = \mu_{C_i}(v)\ and\ \nu_{C_j}(v) = \nu_{C_i}(v)\right)\ then\ (C_i \mapsto C_j) = (C_j \mapsto C_i),$$

and the second

$$if\ \left(\mu_{C_j}(v) = \mu_{C_j}(v)\ and\ \mu_{C_j}(v) = \nu_{C_i}(v)\right)\ then\ (C_i \mapsto C_j) = (C_j \mapsto C_i),$$

and they are obvious from Eqs. 7.3 and 7.4.

Proof Property *(3)*

It can be noticed that for the intuitionistic fuzzy sets $C_i, C_j \in IFS(V)$ the pair of impacts of one intuitionistic fuzzy set on another intuitionistic fuzzy set (according to Eqs. 7.3 and 7.4) is described as follows

$$(C_i \mapsto C_j) = \{(v,\ min\{\mu_{C_i}(v),\ \nu_{C_j}(v)\},\ max\{\nu_{C_i}(v),\ \mu_{C_j}(v)\}) \mid v \in V\},$$

$$(C_j \mapsto C_i) = \{(v,\ min\{\mu_{C_j}(v),\ \nu_{C_i}(v)\},\ max\{\nu_{C_j}(v),\ \mu_{C_i}(v)\}) \mid v \in V\}.$$

The normalized Hamming distance between C_i and C_j takes into account only two parameters, that is, the left hand side of the equality is of the form

$$D_{nHamming}(C_i, C_j) = \frac{1}{2L}\sum_{l=1}^{L}(\mid \mu_{C_i}(v_l) - \mu_{C_j}(v_l) \mid + \mid \nu_{C_i}(v_l) - \nu_{C_j}(v_l) \mid).$$

Let us prove the right hand side of the equality, that is, the one concerning the normalized Hamming distance between $(C_i \mapsto C_j)$ and $(C_j \mapsto C_i)$, which can be written down in the following form

$$D_{nHamming}\big((C_i \mapsto C_j), (C_j \mapsto C_i)\big) =$$

$$= \frac{1}{2L}\sum_{l=1}^{L}(\mid min\{\mu_{C_i}(v_l),\ \nu_{C_j}(v_l)\} - min\{\mu_{C_j}(v_l),\ \nu_{C_i}(v_l)\} \mid +$$

$$+ \mid max\{\nu_{C_i}(v_l),\ \mu_{C_j}(v_l)\} - max\{\nu_{C_j}(v_l),\ \mu_{C_i}(v_l)\} \mid).$$

Additionally, it can be noticed that for the following conditions

$$\mu_{C_i}(v) \leq \nu_{C_j}(v) \text{ and } \nu_{C_i}(v) \geq \mu_{C_j}(v))$$

the normalized Hamming distance between $(C_i \mapsto C_j)$ and $(C_j \mapsto C_i)$ can be written down as follows

$$D_{nHamming}\big((C_i \mapsto C_j), (C_j \mapsto C_i)\big) =$$

$$= \frac{1}{2L}\sum_{l=1}^{L}(\mid \mu_{C_i}(v_l) - \mu_{C_j}(v_l) \mid + \mid \nu_{C_i}(v_l) - \nu_{C_j}(v_l) \mid) = D_{nHamming}(C_i, C_j),$$

and is equal to the normalized Hamming distance between set C_i and set C_j. □

The above described properties of the perturbation of intuitionistic fuzzy sets will be illustrated by the following examples.

Example 7.19 Let us consider the singleton set $V = \{v\}$ and two intuitionistic fuzzy sets $C_1, C_2 \in IFS(V)$, where

$$C_1 = \{(v,\ 0.4,\ 0.4)\} \text{ and } C_2 = \{(v,\ 0.2,\ 0.2)\}.$$

We can obtain the perturbation of the intuitionistic fuzzy set C_2 by the intuitionistic fuzzy set C_1, as

$$(C_1 \mapsto C_2) = \{(v,\ 0.2,\ 0.4)\}.$$

On the other hand, the perturbation of the intuitionistic fuzzy set C_1 by the intuitionistic fuzzy set C_2 is the very same set, namely

$$(C_2 \mapsto C_1) = \{(v,\ 0.2,\ 0.4)\}.$$

□

Example 7.20 Again, let us consider the set $V = \{v\}$, and the intuitionistic fuzzy set

$$C_i = \{(v,\ 0.6,\ 0.3)\},$$

and the complement of this set, denoted by

$$C_i^C = \{(v,\ 0.3,\ 0.6)\},$$

shown as the respective points in Fig. 7.3.

It is interesting that for the fixed intuitionistic fuzzy set $C_i = \{(v, 0.6, 0.3)\}$, and for any intuitionistic fuzzy set C_j from the blue area in Fig. 7.3, the equality

$$D_{nHamming}(C_i, C_j) = D_{nHamming}\big((C_i \mapsto C_j), (C_j \mapsto C_i)\big)$$

is satisfied. □

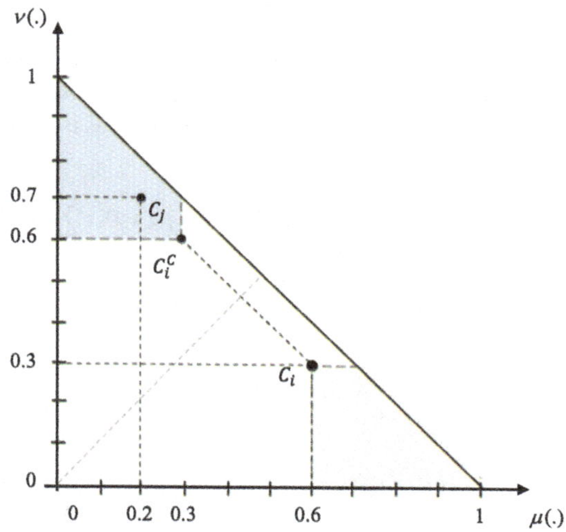

Fig. 7.3 A fixed intuitionistic fuzzy set $C_i = \{(v, 0.6, 0.3)\}$, the complement of this set C_i^C, and an arbitrary intuitionistic fuzzy set C_j from the blue area

7.3.2 Measures of Perturbation of Intuitionistic Fuzzy Sets

There are many different definitions of proximity measures between intuitionistic fuzzy sets. In the case where only the degree of membership $\mu_C(.)$ and non-membership $\nu_C(.)$ are considered, then a two-dimensional representation is used (e.g., [5]). However, in the case where the hesitancy degree $\pi_C(.)$ is also taken into account, then a three-dimensional representation is used (e.g., [6]).

In our approach, we apply the cardinality and the relative cardinality of a finite intuitionistic fuzzy set $C \in IFS(V)$ as a number $card_{IFS}(C)$ and $relcard_{IFS}(C)$, which are constrained to closed intervals, representing the minimum and maximum numbers of elements that the finite intuitionistic fuzzy set may contain. Thus, for any intuitionistic fuzzy set *the cardinality* is expressed as the following interval (see Szmidt and Kacprzyk [8])

$$card_{IFS}(C) = [min\ Count(C),\ max\ Count(C)], \tag{7.5}$$

which can be rewritten in the following form

$$card_{IFS}(C) = \left[\sum_{v \in V} \mu_C(v),\ \sum_{v \in V} (1 - \nu_C(v))\right] = \tag{7.6}$$

$$= \left[\sum_{v \in V} \mu_C(v),\ \sum_{v \in V} (\mu_C(v) + \pi_C(v))\right].$$

The length of such interval corresponds to the uncertainty margin $\pi_C(.)$ of the elements of the intuitionistic fuzzy sets. Note that if the uncertainty margin is equal zero (i.e., for fuzzy sets), then the interval is reduced to a point

$$\left[\sum_{v \in V} \mu_C(v),\ \sum_{v \in V} \mu_C(v)\right].$$

We can also describe *the relative cardinality* of C with respect to V, as follows

$$relcard_{IFS}(C) = \frac{1}{card(V)}\ card_{IFS}(C). \tag{7.7}$$

Thus, the relative cardinality (7.7) is understood as a number from the range established through (7.6), divided by the number of elements L of the set V, as follows

$$relcard_{IFS}(C) = \left[\frac{1}{L}\sum_{v\in V}\mu_C(v),\ \frac{1}{L}\sum_{v\in V}(1-\nu_C(v))\right] = \tag{7.8}$$

$$= \left[\frac{1}{L}\sum_{v\in V}\mu_C(v),\ \frac{1}{L}\sum_{v\in V}(\mu_C(v)+\pi_C(v))\right].$$

Below, we present the measure of perturbation of one intuitionistic fuzzy set C_j by another intuitionistic fuzzy set C_i (or vice versa), defined by a mapping $IFS(V) \times IFS(V) \to [0, 1]$.

Definition 7.4 (*Measure of perturbation of intuitionistic fuzzy set*) Let us assume that $C_i, C_j \in IFS(V)$, and $V = \{v_1, v_2, ..., v_L\}$. The measure of perturbation of one intuitionistic fuzzy set C_j by another intuitionistic fuzzy set C_i, denoted by $Per_{IFS}(C_i \mapsto C_j)$, is defined in the following manner:

$$Per_{IFS}(C_i \mapsto C_j) = relcard_{IFS}(C_i \mapsto C_j) = \tag{7.9}$$

$$= \left[Per_{IFS}^{Left}(C_i \mapsto C_j),\ Per_{IFS}^{Right}(C_i \mapsto C_j)\right], \tag{7.10}$$

where

$$Per_{IFS}^{Left}(C_i \mapsto C_j) = \frac{1}{L}\sum_{l=1}^{L}\mu_{C_i\mapsto C_j}(v_l),$$

$$Per_{IFS}^{Right}(C_i \mapsto C_j) = \frac{1}{L}\sum_{l=1}^{L}(1-\nu_{C_i\mapsto C_j}(v_l)),$$

where the superscripts $Left$ and $Right$ denote the left and right hand ends of the defined interval.

Then, using the definition of perturbation of one intuitionistic fuzzy set by another intutionistic fuzzy set (Definition 7.4) we obtain the following expression for the measure of perturbation of one intuitionistic fuzzy set C_j by another intuitionistic fuzzy set C_i:

$$Per_{IFS}(C_i \mapsto C_j) = \left[\frac{1}{L}\sum_{l=1}^{L}\mu_{C_i\mapsto Cj}(v_l),\ \frac{1}{L}\sum_{l=1}^{L}(1-\nu_{C_i\mapsto C_j}(v_l))\right] =$$

$$= \left[\frac{1}{L}\sum_{l=1}^{L}(min\{\mu_{C_i}(v_l),\ \nu_{C_j}(v_l)\}),\ \frac{1}{L}\sum_{l=1}^{L}(1-max\{\nu_{C_i}(v_l),\ \mu_{C_j}(v_l)\})\right]. \tag{7.11}$$

In the counterpart case, the measure of perturbation of the intuitionistic fuzzy set C_i by the intuitionistic fuzzy set C_j is analogous, and is defined as follows:

$$Per_{IFS}(C_j \mapsto C_i) = \left[\frac{1}{L}\sum_{l=1}^{L}\mu_{C_j \mapsto Ci}(v_l),\ \frac{1}{L}\sum_{l=1}^{L}(1 - \nu_{C_j \mapsto C_i}(v_l))\right] =$$

$$= \left[\frac{1}{L}\sum_{l=1}^{L}(min\{\mu_{C_j}(v_l),\ \nu_{C_i}(v_l)\}),\ \frac{1}{L}\sum_{l=1}^{L}(1 - max\{\nu_{C_j}(v_l),\ \mu_{C_i}(v_l)\})\right]. \tag{7.12}$$

For simplicity and further use, here, we introduce the so called *elementary measure of the perturbation* for a fixed, one value of $v_l \in V$.

Definition 7.5 (*Elementary measure of perturbation*) Let us assume, that $C_i, C_j \in IFS(V)$, and $V = \{v_1, v_2, ..., v_L\}$. The elementary measure of perturbation of one intuitionistic fuzzy set C_j by another intuitionistic fuzzy set C_i, denoted by $Per_{IFS}(C_i \mapsto C_j \mid v_l)$, for value of $v_l \in V$, is defined in the following manner:

$$Per_{IFS}(C_i \mapsto C_j \mid v_l) =$$

$$= \left[Per_{IFS}^{Left}(C_i \mapsto C_j \mid v_l),\ Per_{IFS}^{Right}(C_i \mapsto C_j \mid v_l)\right] =$$

$$= \left[min\{\mu_{C_i}(v_l),\ \nu_{C_j}(v_l)\},\ (1 - max\{\nu_{C_i}(v_l),\ \mu_{C_j}(v_l)\})\right]. \tag{7.13}$$

The above described operation, performed on intuitionistic fuzzy sets, will be illustrated by the following example.

Example 7.21 Let us consider the set $V = \{v\}$ and two intuitionistic fuzzy sets $C_1, C_2 \in IFS(V)$, where

$$C_1 = \{(v,\ 0.2,\ 0.7)\} \text{ and } C_2 = \{(v,\ 0.6,\ 0.1)\}.$$

Then, we can describe the perturbation of the intuitionistic fuzzy set C_2 by the intuitionistic fuzzy set C_1 as follows

$$(C_1 \mapsto C_2) = \{(v,\ 0.1,\ 0.7)\}.$$

On the other hand, the perturbation of the intuitionistic fuzzy set C_1 by the intuitionistic fuzzy set C_2 is given as

$$(C_2 \mapsto C_1) = \{(v,\ 0.6,\ 0.2)\}.$$

According to Eq. 7.13, the measures of the perturbations belong to the following closed intervals:

$$Per_{IFS}(C_1 \mapsto C_2) = [0.1,\ 0.3],$$

$$Per_{IFS}(C_2 \mapsto C_1) = [0.6,\ 0.8].$$

□

7.3.3 Properties of Measure of Perturbation of Intuitionistic Fuzzy Sets

Now, let us state two properties of the measures of perturbation of one intuitionistic fuzzy set by another intuitionistic fuzzy set in the form of the following Corollary.

Corollary 7.23 *Let us consider two intuitionistic fuzzy sets $C_i, C_j \in IFS(V)$, and the measures of perturbation of one intuitionistic fuzzy set by another intuitionistic fuzzy set $Per_{IFS}(C_i \mapsto C_j)$ and $Per_{IFS}(C_j \mapsto C_i)$, respectively. The following properties of the measures of the perturbation hold:*

(1) $Per_{IFS}(C_i \mapsto C_j) \subset [0, 1]$ *and* $Per_{IFS}(C_j \mapsto C_i) \subset [0, 1]$;

(2) $Per_{IFS}(C_i \mapsto C_j) = Per_{IFS}(C_j \mapsto C_i)$ *if and only if* $\big(\mu_{C_j}(v) = \nu_{C_j}(v)$ *and* $\mu_{C_i}(v) = \nu_{C_i}(v)$, *or* $\mu_{C_j}(v) = \mu_{C_i}(v)$ *and* $\nu_{Cj}(v) = \nu_{C_i}(v)\big)$.

Proof *(1)* Obvious by Definition 7.4.

Proof *(2)* Obvious by Corollary 7.22. □

7.3.4 Graphical Illustration of Measures of Perturbation of Intuitionistic Fuzzy Sets

In order to explain the idea of the measure of perturbations of intuitionistic fuzzy sets, let us analyze and discuss its geometric interpretation. Thus, consider the set $V = \{v\}$, and two intuitionistic fuzzy sets C_i, C_j, and the measures of perturbation of the intuitionistic fuzzy sets C_i and C_j (according to (7.10)) belonging to the following intervals

$$Per_{IFS}(C_i \mapsto C_j) = \left[Per_{IFS}^{Left}(C_i \mapsto C_j),\ Per_{IFS}^{Right}(C_i \mapsto C_j)\right],$$

and

$$Per_{IFS}(C_j \mapsto C_i) = \left[Per_{IFS}^{Left}(C_j \mapsto C_i),\ Per_{IFS}^{Right}(C_j \mapsto C_i)\right],$$

where the superscripts $Left$ and $Right$ denote the left and right hand sides of the interval limits, respectively.

The goal is to determine the mutual positions of these intervals. For this purpose, let us consider the difference between the right hand side of the measure of perturbation $(C_i \mapsto C_j)$ and the left hand side of the measure of perturbation $(C_j \mapsto C_i)$. According to (7.11) and (7.12), we obtain the following relationship

$$Per_{IFS}^{Right}(C_i \mapsto C_j) - Per_{IFS}^{Left}(C_j \mapsto C_i) =$$

$$= 1 - max\{\nu_{C_i}(v), \mu_{C_j}(v)\} - min\{\mu_{C_j}(v), \nu_{C_i}(v)\} =$$

$$= 1 - \left(\mu_{C_j}(v) + \nu_{C_i}(v)\right). \tag{7.14}$$

Let us consider three possible cases leading to different values in Eq. 7.14:

$$Case\ 1: \ \mu_{C_j}(v) + \nu_{C_i}(v) = 1,$$

$$Case\ 2: \ \mu_{C_j}(v) + \nu_{C_i}(v) > 1,$$

$$Case\ 3: \ \mu_{C_j}(v) + \nu_{C_i}(v) < 1.$$

The possible arrangements of the considered intervals relative to each other and the corresponding examples are shown below. Additionally, for these three cases, the following selected distances: the normalized Hamming distance $D_{nHamming}(C_i, C_j)$ between two intuitionistic fuzzy sets C_i and C_j, and the normalized Euclidean distance $D_{nEuclidean}(C_i, C_j)$ between two intuitionistic fuzzy sets C_i and C_j were calculated (for details see Sect. 7.3.2).

- *Case* 1

If the equality $\mu_{C_j}(v) + \nu_{C_i}(v) = 1$ holds, then the considered intervals are directly adjacent.

For the example of the intuitionistic fuzzy sets:

$$C_i = \{(v, 0.2, 0.4)\} \text{ and } C_j = \{(v, 0.6, 0.1)\},$$

and the perturbations

$$(C_i \mapsto C_j) = \{(v, 0.1, 0.6)\},$$

$$(C_j \mapsto C_i) = \{(v, 0.4, 0.2)\},$$

the measures of perturbation

$$Per_{IFS}(C_i \mapsto C_j) = [0.1, 0.4],$$

$$Per_{IFS}(C_j \mapsto C_i) = [0.4, 0.8],$$

are depicted in Fig. 7.4.

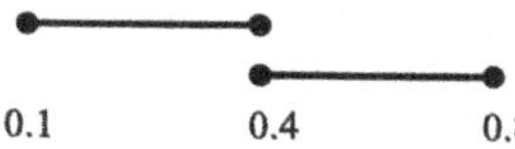

Fig. 7.4 Directly adjacent intervals, $\mu_{C_j}(v) + \nu_{C_i}(v) = 1$

Fig. 7.5 Separate intervals, $\mu_{C_j}(v) + \nu_{C_i}(v) > 1$

The values of the normalized Hamming distance and of the normalized Euclidean distance are calculated as follows:

$$D_{nHamming}(C_i, C_j) = 0.35 \quad \text{and} \quad D_{nEuclidean}(C_i, C_j) = 0.5.$$

- *Case* 2

If the inequality $\mu_{C_j}(v) + \nu_{C_i}(v) > 1$ holds, then the considered intervals are separated.

For the example of intuitionistic fuzzy sets:

$$C_i = \{(v, 0.2, 0.7)\} \text{ and } C_j = \{(v, 0.6, 0.1)\},$$

and the perturbation

$$(C_i \mapsto C_j) = \{(v, 0.1, 0.7)\},$$

$$(C_j \mapsto C_i) = \{(v, 0.6, 0.2)\},$$

the measures of perturbation

$$Per_{IFS}(C_i \mapsto C_j) = [0.1, 0.3]\,,$$

$$Per_{IFS}(C_j \mapsto C_i) = [0.6, 0.8]$$

are shown in Fig. 7.5.

The values of the normalized Hamming distance and of the normalized Euclidean distance are as follows:

$$D_{nHamming}(C_i, C_j) = 0.5 \quad \text{and} \quad D_{nEuclidean}(C_i, C_j) = 0.72.$$

- *Case* 3

If the inequality $\mu_{C_j}(v) + \nu_{C_i}(v) < 1$ holds, then the corresponding intervals are overlapping.

Again, for the example of the intuitionistic fuzzy sets:

$$C_i = \{(v, 0.4, 0.3)\} \text{ and } C_j = \{(v, 0.6, 0.1)\}.$$

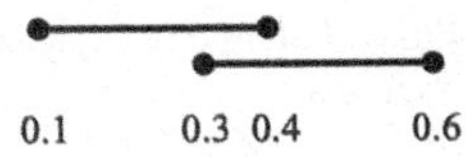

Fig. 7.6 Overlapping intervals, $\mu_{C_j}(v) + \nu_{C_i}(v) < 1$

and the perturbation

$$(C_i \mapsto C_j) = \{(v, 0.1, 0.6)\},$$

$$(C_j \mapsto C_i) = \{(v, 0.3, 0.4)\},$$

the measures of perturbation

$$Per_{IFS}(C_i \mapsto C_j) = [0.1, 0.4]\,,$$

$$Per_{IFS}(C_j \mapsto C_i) = [0.3, 0.6]\,,$$

are depicted in Fig. 7.6.

In this case the values of the normalized Hamming distance and of the normalized Euclidean distance are as follows:

$$D_{nHamming}(C_i, C_j) = 0.2 \quad \text{and} \quad D_{nEuclidean}(C_i, C_j) = 0.2.$$

7.4 Representation of Multi-attribute Objects by Intuitionistic Fuzzy Sets

Now, we will consider objects described by intuitionistic fuzzy sets. Namely, we have a non-empty, finite set of objects U, where

$$U = \{e_1, e_2, ..., e_N\},$$

described by K nominal attributes

$$A = \{a_1, a_2, ..., a_K\}.$$

The finite set

$$V_{a_j} = \{v_{j,1}, v_{j,2}, ..., v_{j,L_j}\}$$

is a domain of the attribute a_j, for $j = 1, 2, ..., K$, and

$$V = \bigcup_{a_j \in A} V_{a_j}.$$

Then, let us suppose that the value

$$v_{j,t(j,n)} \in V_{a_j}$$

of the attribute a_j, for $j = 1, 2, ..., K$, is not precisely known for the object $e_n \in U$. The index

$$t(j, n), \quad \text{for} \quad j \in \{1, 2, ..., K\} \quad \text{and} \quad n \in \{1, 2, ..., N\},$$

specifies which value of the j-th attribute occurs in the n-th object.

The following information is available:

- *positive* - the value is possible or acceptable in some degree, and
- *negative* - this value is impossible or unacceptable in some degree.

Thus, such information can be represented by two degree values (i.e., one degree of positiveness and the other of negativeness), denoted by

$$\mu(e_n, v_{j,t(j,n)}) \text{ and } \nu(e_n, v_{j,t(j,n)})$$

on V_{a_j}, for $j = 1, 2, ..., K$. The value

$$\mu(e_n, v_{j,t(j,n)})$$

establishes, that $v_{j,t(j,n)}$ represents actually the possible value of attribute a_j for object e_n. On the other hand, the value

$$\nu(e_n, v_{j,t(j,n)})$$

specifies the degree, to which the value of the attribute a_j cannot be treated as acceptable for object e_n.

Usually, the values of both positive and negative information are restricted as follows:

$$0 \leq \mu(e_n, v_{j,t(j,n)}) \leq 1,$$

$$0 \leq \nu(e_n, v_{j,t(j,n)}) \leq 1.$$

Now, let us assume that the sum of such degrees for object e_n and the fixed value $v_{j,t(j,n)} \in V_{a_j}$ satisfies the following inequality (one of the assumptions of the intuitionistic fuzzy set modeling)

$$0 \leq \mu(e_n, v_{j,t(j,n)}) + \nu(e_n, v_{j,t(j,n)}) \leq 1, \tag{7.15}$$

where

$$\mu(e_n, v_{j,t(j,n)}) \in [0, 1] \text{ and } \nu(e_n, v_{j,t(j,n)}) \in [0, 1].$$

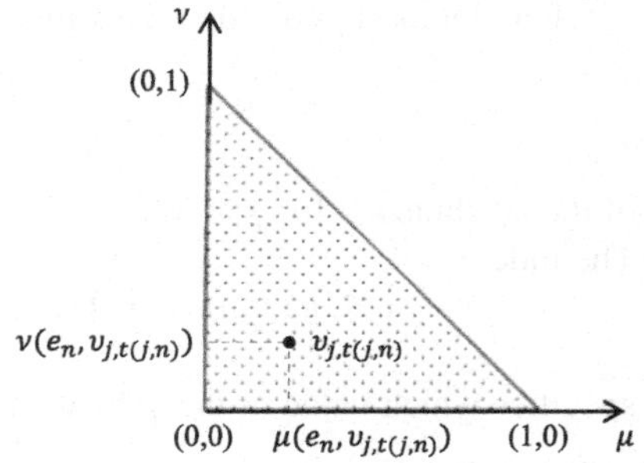

Fig. 7.7 Geometric representation of the element $\upsilon_{j,t(j,n)} \in V_{a_j}, e_n \in U$

Therefore, we can assume that the object $e_n \in U$ for any element $\upsilon_{j,t(j,n)} \in V_{a_j}$, for $j = 1, 2, ..., K$, is described via two coordinates

$$\left(\mu(e_n, \upsilon_{j,t(j,n)}),\ \nu(e_n, \upsilon_{j,t(j,n)})\right),$$

which can be represented by a point in the triangle, as shown in Fig. 7.7.

The value

$$\pi(e_n, \upsilon_{j,t(j,n)}) = 1 - \mu(e_n, \upsilon_{j,t(j,n)}) - \nu(e_n, \upsilon_{j,t(j,n)})$$

can be treated as *the degree of indeterminacy* of the element $\upsilon_{j,t(j,n)}$ for the object e_n.

Now, let us assume that the object $e_n \in U$ and the value $\upsilon_{j,t(j,n)} \in V_{a_j}$ represent the actual value of attribute a_j for object e_n. Thereby, we can define *the intuitionistic selector* in the following way

$$\left((a_j = \upsilon_{j,t(j,n)}),\ \mu(e_n, \upsilon_{j,t(j,n)}),\ \nu(e_n, \upsilon_{j,t(j,n)})\right), \tag{7.16}$$

for the j-th attribute and the object e_n.

For simplicity, we use below the following short notation for the intuitionistic selector

$$\left((a_j = \upsilon_{j,t(j,n)}),\ \mu_{j,t(j,n)},\ \nu_{j,t(j,n)}\right), \tag{7.17}$$

or

$$\left(\upsilon_{j,t(j,n)},\ \mu_{j,t(j,n)},\ \nu_{j,t(j,n)}\right).$$

The index $t(j, n)$, is the same as previously, and specifies that value $\upsilon_{j,t(j,n)}$ of the j-th attribute appearing in the intuitionistic selector.

The function $\mu_{j,t(j,n)}$ specifies that $\upsilon_{j,t(j,n)}$ represents the actually possible value of the attribute a_j for the object e_n. The condition $\mu_{j,t(j,n)} = 0$ means that there is no such confirmation that the value $\upsilon_{j,t(j,n)}$ represents the attribute a_j; while the condition $\mu_{j,t(j,n)} = 1$ means that the value $\upsilon_{j,t(j,n)}$ represents for sure the attribute

a_j for object e_n (i.e., fully confirmed, the value 1 is the largest positive value on the positive scale).

The function $\nu_{j,t(j,n)}$ determines the degree, to which the value $v_{j,t(j,n)}$ of the attribute a_j for the object e_n cannot be acceptable. Thus, the condition $\nu_{j,t(j,n)} = 1$ means that it is totally impossible to accept the value $v_{j,t(j,n)}$ of the attribute a_j for the object e_n. On the other hand, the condition $\nu_{j,t(j,n)} = 0$ means that it is potentially fully possible to accept the value $v_{j,t(j,n)}$ of the attribute a_j for the object. Thus, the interval [0,1] is the considered scale.

In order to clarify the above discussion about the degrees of acceptance and not acceptance of values of attributes of objects let us consider the following illustrative example.

Example 7.22 For the object $e_1 \in U$, and for the attribute $a_1 = Age$, and for the values of this attribute

$$V_{a_1} = \{v_{1,1},\ v_{1,2},\ v_{1,3}\} = \{young, middle, old\}$$

(according to Eq. 7.16), the attribute Age of the object $person\#1$ can be described by the following exemplary intuitionistic selector

$$\big((a_1 = v_{1,2}),\ \mu_{1,2},\ \nu_{1,2}\big) = \big((Age = middle),\ 0.75,\ 0.1\big).$$

□

Let us again consider an object described by two coordinates. Thus, we assume that for an object $e_n \in U$, each value in the domain of the attributes (values $v_{j,t(j,n)} \in V_{a_j}$, for $j = 1, 2, ..., K$), is described via two coordinates

$$(\mu(e_n, v_{j,t(j,n)}),\ \nu(e_n, v_{j,t(j,n)})).$$

Then, let us consider the intuitionistic selector for which the condition

$$0 \le \nu(e_n, v_{j,t(j,n)}) \le \mu(e_n, v_{j,t(j,n)}) \le 1 \tag{7.18}$$

is satisfied.

For example, the intuitionistic selector $\big((Age = middle),\ 0.75,\ 0.1\big)$ from Example 7.22 satisfies the condition (7.18).

Thus, the selected element $v_{j,t(j,n)}$ can be represented by a point in the respective triangle, see Fig. 7.8.

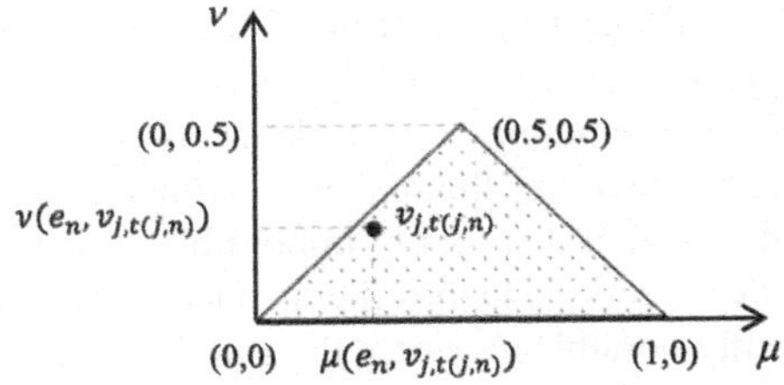

Fig. 7.8 Geometric representation of the element $\upsilon_{j,t(j,n)} \in V_{a_j}$, satisfying the condition (7.18), $e_n \in U$

For convenience, a simplified description of objects is applied, namely the term $(e_n, \upsilon_{j,t(j,n)})$ is skipped. Therefore, we often use the designation

$$\mu_{j,t(j,n)} \text{ instead of } \mu(e_n, \upsilon_{j,t(j,n)}),$$

and

$$\nu_{j,t(j,n)} \text{ instead of } \nu(e_n, \upsilon_{j,t(j,n)}).$$

Thus, according to the introduced notation, the object e_n is constituted by the conjunction of K intuitionistic selectors, in the following way

$$e_n = \left((a_1 = \upsilon_{1,t(1,n)}),\ \mu_{1,t(1,n)},\ \nu_{1,t(1,n)}\right) \wedge \ldots$$

$$\ldots \wedge \left((a_K = \upsilon_{K,t(K,n)}),\ \mu_{K,t(K,n)},\ \nu_{K,t(K,n)}\right), \tag{7.19}$$

for any $j = 1, 2, \ldots, K$. For the values of each attribute belonging to predefined domains, $\upsilon_{j,t(j,n)} \in V_{a_j}$, the values of membership and non-membership (i.e., $\mu_{j,t(j,n)}, \nu_{j,t(j,n)} \in [0, 1]$), satisfy the conditions

$$0 \leq \nu_{j,t(j,n)} \leq \mu_{j,t(j,n)} \leq 1.$$

The above considerations are based on the assumption that each object is described by only one value of each attribute. The extension, meaning that each object can be described by all the values of each attribute, will be described in a similar way, however more complex.

Now, let us consider groups of objects, g_1 and g_2, represented as intuitionistic fuzzy sets. The descriptions of such groups (denoted by G_{g_1} and G_{g_2}, respectively) can be created, for example, through application of the average of a collection of intuitionistic fuzzy sets in the group. Thereby, the descriptions of the groups, G_{g_1} and G_{g_2}, are also represented by the intuitionistic fuzzy sets.

For simplicity of notation and further use (see Sect. 7.5.2) we will limit ourselves to the case when $K = 1$ (i.e., to one attribute a_1) with the domain $V_{a_1} = \{\upsilon_1, \upsilon_2, \ldots, \upsilon_L\}$.

We consider the case, when each object is described by all values from the domain V_{a_1}.

Thus, *the elementary perturbation*, denoted by $(G_{g_1} \mapsto G_{g_2} \mid v_l)$, can be constructed, for $l = 1, 2, ..., L$, and then *the elementary measure of perturbation*, denoted by $Per_{IFS}(G_{g_1} \mapsto G_{g_2} \mid v_l)$, can be calculated (according to Eq. 7.13), for the actual value $v_l \in V_{a_1}$, as follows

$$Per_{IFS}(G_{g_1} \mapsto G_{g_2} \mid v_l) =$$

$$= \left[Per_{IFS}^{Left}(G_{g_1} \mapsto G_{g_2} \mid v_l),\ Per_{IFS}^{Right}(G_{g_1} \mapsto G_{g_2} \mid v_l)\right] =$$

$$= \left[\mu_{(G_{g_1} \mapsto G_{g_2} \mid v_l)}(v_l),\ (1 - \nu_{(G_{g_1} \mapsto G_{g_2} \mid v_l)}(v_l))\right].$$

A practical application of the elementary measure of perturbation (with assumptions as specified above) will be shown in Sect. 7.5.2.

7.5 Illustrative Examples

Before the detailed descriptions of the applications let us introduce the following three definitions, which are necessary and useful for comparing the measures of perturbation of the intuitionistic fuzzy sets.

The first definition concerns the conversion of an interval into a proper intuitionistic fuzzy set.

Definition 7.6 (*Conversion of interval into intuitionistic fuzzy set*) Let us consider an interval $[a, b]$, for $a \leq b$, where $a, b \in [0, 1]$. The conversion operation, denoted by $C_{\rightleftharpoons}(.)$, converts the interval to an ordered pair of numbers in the following way

$$C_{\rightleftharpoons}([a, b]) = (c, d) \tag{7.20}$$

where $c := a$ and $d := b - a$.

Next, such an ordered pair of numbers (c, d), for a fixed $v \in V$, can be converted into an intuitionistic fuzzy set $\{(v, \mu(v), \nu(v))\}$ in the following way:

$$\mu(.) := c \quad and \quad \nu(.) := 1 - \mu(.) - \pi(.) \tag{7.21}$$

where $\pi(.) := d$.

The process of conversion from an interval into an intuitionistic fuzzy set is shown in the following short example. Namely, according to Definition 7.6, the exemplary interval

$$[0.57, 0.67]$$

can be converted into the ordered pair of numbers

$$(0.57, 0.1),$$

and next to the intuitionistic fuzzy set

$$\{(v, 0.57, 0.33)\}.$$

The following definitions specifies the conditions for two pairs of numbers to create a lexicographical ordering, namely.

Definition 7.7 (*Sharp lexicographical ordering* $\prec_L$) Let us consider the ordered pairs $(a, b), (a', b') \in [0, 1]^2$. A sharp lexicographical ordering $\prec_L$ on the ordered pairs of numbers is defined as

$$(a, b) \prec_L (a', b') \ if \ and \ only \ if \quad a < a' \ or \ (a = a' \ and \ \ b < b'), \tag{7.22}$$

where $<$ is the usual order relation.

The above definition determines whether the pair (a, b) occurs in the ordering earlier or later than the pair (a', b'), creating a sharp lexicographical ordering. At the beginning, the first elements of each pair are compared, and, if need arises, then the second elements of each pair are compared.

Definition 7.8 (*Lexicographical ordering* $\preceq_L$) Let us consider the ordered pairs $(a, b), (a', b') \in [0, 1]^2$. A lexicographical ordering $\preceq_L$ on the ordered pairs of numbers is defined as

$$(a, b) \preceq_L (a', b') \ if \ and \ only \ if \ \big((a, b) \prec_L (a', b') \ or \ (a, b) = (a', b')\big), \tag{7.23}$$

where

$$(a, b) = (a', b') \ if \ and \ only \ if \ \ (a = a' \ and \ \ b = b').$$

In order to bring closer the idea, how the lexicographical ordering is realized, let us discuss the following illustrative example.

7.5.1 Pattern Matching

The first application is based on lexicographical ordering of pairs of numbers or otherwise on the ordering of intuitionistic fuzzy sets.

Let us consider three patterns, expressed by the intuitionistic fuzzy sets $C_1, C_2, C_3 \in IFS(V)$, in the form

$$C_1 = \{(0.4, 0.5), (0.7, 0.1), (0.3, 0.3)\},$$

$$C_2 = \{(0.5, 0.4), (0.7, 0.2), (0.4, 0.3)\},$$

$$C_3 = \{(0.4, 0.5), (0.7, 0.1), (0.4, 0.3)\},$$

and example another pattern B as the intuitionistic fuzzy set $B \in IFS(V)$, taken from the work [9], as follows

$$B = \{(0.1,\ 0.1), (1.0,\ 0.0), (0.0,\ 1.0)\}.$$

The problem is to find the set C_i, $i \in \{1, 2, 3\}$, to which the set B is most similar.

First, let us calculate the normalized Hamming distance and the normalized Euclidean distance for these sets [9]. It is easy to see that in this case all the considered normalized Hamming distances are equal

$$D_{nHamming}(C_1, B) = D_{nHamming}(C_2, B) = D_{nHamming}(C_3, B) = 0.483,$$

and the normalized Euclidean distances are also all equal

$$D_{nEuclidean}(C_1, B) = D_{nEuclidean}(C_2, B) = D_{nEuclidean}(C_3, B) = 0.442.$$

Thus, the same values were obtained, therefore these distance measure do not differentiate the tested distances.

Next, let us recall the distance suggested in [9], the obtained values of distances are as follows:

$$D(C_1, B) = 0.401, \quad D(C_2, B) = 0.418, \quad D(C_3, B) = 0.442.$$

In the light the above values, the set B is the closest to the set C_1 [9].

Now, let us apply the measure of the perturbation of intuitionistic fuzzy sets, developed in this book. Following Eq. 7.12, we calculate the measures of the following perturbations

$$Per_{IFS}(B \mapsto C_1), \quad Per_{IFS}(B \mapsto C_2), \quad \text{and} \quad Per_{IFS}(B \mapsto C_3).$$

The results are shown in Table 7.1, in the second column. In the third column of Table 7.1, conversions of the measures of perturbation of the intuitionistic fuzzy sets into corresponding pairs of numbers are shown, according to Eq. 7.20.

When examining the third column in Table 7.1, we can notice that according to Eq. 7.23, the lexicographical ordering of the measures of perturbation is as follows

$$Per_{IFS}(B \mapsto C_1) \preceq_L Per_{IFS}(B \mapsto C_2) \preceq_L Per_{IFS}(B \mapsto C_3).$$

Thus, according to the obtained lexicographical ordering of the pairs (i.e., column 3 in Table 7.1) the set B is the closest to the set C_1, because the respective measure of perturbation is the smallest.

In fact, the approach based on the perturbation of intuitionistic fuzzy sets confirms the results obtained in the paper by Zhang and Yu [9].

7.5.2 Generation of Classification Rules

Let us consider the set of objects represented as intuitionistic fuzzy sets. Next, let us assume that the considered objects are divided into two separate groups, namely g_1 and g_2. The aim is to construct the classification rules for the distinguished groups of objects.

In this section, the generation of such rules is performed on the basis of perturbations of the intuitionistic fuzzy sets. The rules allow for distinguishing the considered group from the rest of objects, namely those belonging to other group. The classification rules are generated separately for each group (cf. Kacprzyk and Szkatuła [10]). Each such classification rule has the well known form:

$$IF\ certain\ conditions\ are\ satisfied$$

$$THEN\ a\ given\ object\ is\ a\ member\ of\ a\ specific\ group.$$

Such generated classification rules can also be applied to classify new objects. The classification is carried out through verification of fulfilment of conditions in the conditional parts of the rules.

The respective algorithm can be roughly described in the following steps.

Step 1

Two groups of objects g_1 and g_2 are considered, and the objects are represented as the intuitionistic fuzzy sets.

Table 7.1 Conversion of the measures of perturbation into corresponding pairs of numbers

	$Per_{IFS}(.)$	$C_{=}(Per_{IFS}(.))$
$(B \mapsto C_1)$	$[0.067,\ 0.3]$	$(0.067,\ 0.233)$
$(B \mapsto C_2)$	$[0.1,\ 0.267]$	$(0.1,\ 0.167)$
$(B \mapsto C_3)$	$[0.17,\ 0.3]$	$(0.17,\ 0.13)$

Step 2

The descriptions of the groups of objects (denoted by G_{g_1} and G_{g_2}) are obtained, by applying, for example, the average of a collection of IFS sets in each group. So, the descriptions of the groups are also constituted by the intuitionistic fuzzy sets.

Step 3

Separately, for each value $v_l \in V$, the elementary perturbation $(G_{g_1} \mapsto G_{g_2} \mid v_l)$ is constructed, as well as the elementary measure of the perturbation of one intuitionistic fuzzy set by another, according to Eq. 7.13. Each obtained interval corresponds to the uncertainty margin.

Step 4

The obtained intervals are converted to ordered pairs of numbers, according to Eq. 7.20. Then, the lexicographical ordering of the ordered pairs is calculated, according to Eqs. 7.22 and 7.23.

Step 5

A real number, being a parameter $\alpha \in [0, 1]$, treated as the $\alpha - threshold$, is introduced. In this step, we perform the reduction of the ordered pairs of numbers. The reduction is done via consideration of only these pairs of numbers, in which the first values are greater than or equal to the value of the parameter α.

Step 6

Now, the ordered pairs of numbers are converted to the intuitionistic fuzzy sets (according to Eq. 7.21), which can be used to create the set of elementary rules describing the group g_1.

The above procedure is repeated to create the classification rule for the other considered group.

Practical presentation of the proposed approach will be done for the task of grouping of the text documents, assuming that the context and the semantics can be neglected. We use the examples of the documents, described by the multisets, published in an authors' earlier paper [11]. In this section, we use the data converted into the intuitionistic fuzzy sets representation, for which the classification rules for the groups were generated (cf. [7]). The details of procedure applied are provided below.

- *Data processing*

Let us assume that the sample consists of objects being text documents e_n, for $n = 1, 2, ..., 10$, which are modeled by the multisets S_{e_n} drawn from the set V of the keywords and phrases appearing in the text, where $V = \{v_1, v_2, ..., v_6\}$, and each document can be represented by a set of six-ordered pairs [11] (details can be found in Sect. 5.6.4). Usually, the keywords and phrases can be weighted in various ways, but here, for simplicity, we assume the same importance of any keyword.

The text documents, modeled by multisets S_{e_n}, were converted to the intuitionistic fuzzy sets C_{e_n}, drawn from the set of unique keywords and phrases appearing

Table 7.2 Intuitionistic fuzzy sets C_{e_n}, $n = 1, 2, ..., 10$, representing the text documents

n	v_1	v_2	v_3	v_4	v_5	v_6
1	(0.71, 0.19)	(0.65, 0.25)	(0.53, 0.37)	(0.28, 0.62)	(0.28, 0.62)	(0.24, 0.66)
2	(0.27, 0.63)	(0.25, 0.65)	(0.26, 0.64)	(0.69, 0.21)	(0.69, 0.21)	(0.59, 0.31)
3	(0.27, 0.63)	(0.65, 0.25)	(0.26, 0.64)	(0.28, 0.62)	(0.28, 0.62)	(0.69, 0.21)
4	(0.58, 0.32)	(0.25, 0.65)	(0.66, 0.24)	(0.69, 0.21)	(0.28, 0.62)	(0.35, 0.55)
5	(0.27, 0.63)	(0.25, 0.65)	(0.26, 0.64)	(0.69, 0.21)	(0.69, 0.21)	(0.47, 0.43)
6	(0.42, 0.48)	(0.65, 0.25)	(0.53, 0.37)	(0.28, 0.62)	(0.28, 0.62)	(0.24, 0.66)
7	(0.71, 0.19)	(0.25, 0.65)	(0.66, 0.24)	(0.28, 0.62)	(0.28, 0.62)	(0.24, 0.66)
8	(0.27, 0.63)	(0.65, 0.25)	(0.26, 0.64)	(0.69, 0.21)	(0.69, 0.21)	(0.69, 0.21)
9	(0.71, 0.19)	(0.65, 0.25)	(0.75, 0.15)	(0.28, 0.62)	(0.28, 0.62)	(0.24, 0.66)
10	(0.27, 0.63)	(0.25, 0.65)	(0.26, 0.64)	(0.28, 0.62)	(0.69, 0.21)	(0.69, 0.21)

in the text, $V = \{v_1, v_2, ..., v_6\}$, according to the work [12], and can thus each be represented by a set of six triples of the form

$$\{(\text{keyword or phrase in the text document}, \\ \text{degree of membership}, \ \text{degree of non- membership})\},$$

see Table 7.2.

Next, the text documents were divided into two separated groups, according to the division from work [11], namely

$$g_1 = \{e_1, e_4, e_6, e_7, e_9\} \text{ and } g_2 = \{e_2, e_3, e_5, e_8, e_{10}\}.$$

The aim is to generate the set of elementary rules for the classification of the considered text documents into the two groups considered: g_1 or g_2.

At the beginning, the data set is split into two sets:

- *the learning set* (i.e., the first three text documents from each group),
- *the testing set* (i.e., other text documents).

Next, based on the learning set, the classification rules are generated. The testing set contains other text documents, which do not participate in the generation of the rules and are used to check the quality of the generated classification rules.

The details of the algorithm are described in the following steps.

- *Generation of the classification rules for group* g_1

Step 1

Let us consider *the learning set*, that is—six text documents from two groups of objects, as follows

$$\{e_1, e_4, e_6\} \text{ belonging to the group } g_1,$$

Table 7.3 Descriptions of the groups g_1 and g_2

Description	v_1	v_2	v_3	v_4	v_5	v_6
G_{g_1}	(0.57, 0.33)	(0.52, 0.38)	(0.58, 0.32)	(0.42, 0.48)	(0.28, 0.62)	(0.28, 0.62)
G_{g_2}	(0.27, 0.63)	(0.38, 0.52)	(0.26, 0.64)	(0.55, 0.35)	(0.55, 0.35)	(0.59, 0.31)

Table 7.4 The elementary perturbations $(G_{g_1} \mapsto G_{g_2} \mid v_l)$, $l = 1, 2, ..., 6$, converted to ordered pairs of numbers

v_l	$(G_{g_1} \mapsto G_{g_2} \mid v_l)$	$Per_{IFS}(G_{g_1} \mapsto G_{g_2} \mid v_l)$	$C_{\Leftarrow}(Per_{IFS}(G_{g_1} \mapsto G_{g_2} \mid v_l))$
v_1	(0.57, 0.33)	[0.57, 0.67]	**(0.57, 0.1)**
v_2	(0.52, 0.38)	[0.52, 0.62]	(0.52, 0.1)
v_3	(0.58, 0.32)	[0.58, 0.68]	**(0.58, 0.1)**
v_4	(0.35, 0.55)	[0.35, 0.45]	(0.35, 0.1)
v_5	(0.28, 0.62)	[0.28, 0.38]	(0.28, 0.1)
v_6	(0.28, 0.62)	[0.28, 0.38]	(0.28, 0.1)

and

$$\{e_2, e_3, e_5\} \text{ belonging to the group } g_2,$$

represented as the intuitionistic fuzzy sets, see Table 7.2.

Step 2

Let us form the descriptions of the groups g_1 and g_2 (denoted by G_{g_1} and G_{g_2}, respectively), by applying, for example, the average of the collection of IFSs learning sets in the groups.

In this manner, the descriptions of the groups, G_{g_1} and G_{g_2}, are represented by the intuitionistic fuzzy sets, as shown in Table 7.3.

Step 3

Separately, for each value v_l, $l = 1, 2, ..., 6$ (the first column in Table 7.4), the elementary perturbation $(G_{g_1} \mapsto G_{g_2} \mid v_l)$ is constructed (see the second column in Table 7.4), and the elementary measure of the perturbation of one intuitionistic fuzzy set by another is calculated (according to Eq. 7.13) - see the third column in Table 7.4.

$$Per_{IFS}(G_{g_1} \mapsto G_{g_2} \mid v_l) =$$

$$= \left[Per_{IFS}^{Left}(G_{g_1} \mapsto G_{g_2} \mid v_l),\ Per_{IFS}^{Right}(G_{g_1} \mapsto G_{g_2} \mid v_l) \right] =$$

$$= \left[\mu_{(G_{g_1} \mapsto G_{g_2} | v_l)}(v_l),\ (1 - \nu_{(G_{g_1} \mapsto G_{g_2} | v_l)}(v_l)) \right].$$

Step 4

Each interval from the third column of Table 7.4 is converted to the corresponding ordered pair of numbers, using the operation $C_{\rightleftharpoons}(.)$ according to Eq. 7.20, see the fourth column of Table 7.4.

Next, the lexicographical ordering on the ordered pairs of numbers is calculated according to Eqs. 7.22 and 7.23.

For simplicity, we denote

$$Per_{IFS}(G_{g_1} \mapsto G_{g_2} \mid v_l) \text{ by } Per_{IFS}(v_l).$$

The result is as follows:

$$Per_{IFS}(v_6) \preceq_L Per_{IFS}(v_5) \prec_L Per_{IFS}(v_4) \prec_L Per_{IFS}(v_2) \prec_L$$

$$\prec_L Per_{IFS}(v_1) \prec_L Per_{IFS}(v_3)$$

that is

$$(0.28, 0.1) \preceq_L (0.28, 0.1) \prec_L (0.35, 0.1) \prec_L (0.52, 0.1) \prec_L$$

$$\prec_L \mathbf{(0.57, 0.1)} \prec_L \mathbf{(0.58, 0.1)}.$$

Step 5

We assume, for example, that $\alpha = 0.55$, and the reduction is done via consideration of only those pairs, in which the first values are greater than or equal to the value 0.55 (marked in bold in Table 7.4). Now, after reduction, we consider the following pair of numbers

$$Per_{IFS}(v_1) \prec_L Per_{IFS}(v_3),$$

that is

$$\mathbf{(0.57, 0.1)} \prec_L \mathbf{(0.58, 0.1)}.$$

Step 6

The ordered pairs of numbers

$$(0.57, 0.1) \text{ and } (0.58, 0.1)$$

are converted, according to Eq. 7.21, to the intuitionistic fuzzy sets

$$\{(v_1, 0.57, 0.33)\} \text{ and } \{(v_3, 0.58, 0.32)\},$$

respectively. According to the definition of the intuitionistic selector (7.17), the following classification rule was generated for describing the group g_1.

$$R_{g_1}^{0.55} : IF \ \big((value = v_3), 0.58, 0.32\big) \vee \big((value = v_1), 0.57, 0.33\big)$$

$$THEN \ a \ given \ object \ is \ a \ member \ of \ the \ group \ g_1.$$

The generated rule consists of two elementary selectors, one related to the attribute value v_3 and the second to the attribute value v_1.

It must be emphasized that the same data were already considered, being first represented by multisets (see Sect. 5.6.4), while in this chapter they are represented by intuitionistic fuzzy sets. For these two cases, the set perturbation concept was applied. In the first case it was the perturbation of one multiset by another multiset, while in the second case - the perturbation of one intuitionistic fuzzy set by another intuitionistic fuzzy set. In both cases the generated classification rules use the same attribute values v_1 and v_3 to select the objects belonging to group g_1 from the considered data.

- *Generation of the classification rule for group* g_2

Step 1 and *Step* 2
are the same as those for generating the rule for the group g_1.

Step 3
Separately, for each value v_l, where $l = 1, 2, ..., 6$, the elementary perturbation $(G_{g_2} \mapsto G_{g_1} \mid v_l)$ is constructed (see the second column in Table 7.5). Then, the elementary measure of perturbation of one intuitionistic fuzzy set by another is constructed, according to Eq. 7.13 (see the third column in Table 7.5).

$$Per_{IFS}(G_{g_2} \mapsto G_{g_1} \mid v_l) =$$

$$= \Big[Per_{IFS}^{Left}(G_{g_2} \mapsto G_{g_1} \mid v_l), \ Per_{IFS}^{Right}(G_{g_2} \mapsto G_{g_1} \mid v_l)\Big] =$$

$$= \big[\mu_{(G_{g_2} \mapsto G_{g_1} \mid v_l)}(v_l), \ (1 - \nu_{(G_{g_2} \mapsto G_{g_1} \mid v_l)}(v_l))\big].$$

Step 4
The obtained intervals are converted, using the operation $C_{\rightleftharpoons}(.)$, to the ordered pairs of numbers, according to Eq. 7.20 (see the fourth column in Table 7.5).
For simplicity, we denote

$$Per_{IFS}(G_{g_2} \mapsto G_{g_1} \mid v_l) \ \text{by} \ Per_{IFS}(v_l).$$

Next, the lexicographical ordering on the ordered pairs of numbers is calculated according to Eqs. 7.22 and 7.23, and the result is as follows:

$$Per_{IFS}(v_3) \prec_L Per_{IFS}(v_1) \prec_L Per_{IFS}(v_2) \prec_L Per_{IFS}(v_4) \prec_L$$

$$\prec_L Per_{IFS}(v_5) \prec_L Per_{IFS}(v_6).$$

Table 7.5 The elementary perturbations $(G_{g_2} \mapsto G_{g_1} \mid v_l)$, $l = 1, 2, ..., 6$, converted to ordered pairs of numbers

v_l	$(G_{g_2} \mapsto G_{g_1} \mid v_l)$	$Per_{IFS}(G_{g_2} \mapsto G_{g_1} \mid v_l)$	$C_{\leftrightarrows}(Per_{IFS}(G_{g_2} \mapsto G_{g_1} \mid v_l))$
v_1	(0.27, 0.63)	[0.27, 0.37]	(0.27, 0.1)
v_2	(0.38, 0.52)	[0.38, 0.48]	(0.38, 0.1)
v_3	(0.26, 0.64)	[0.26, 0.36]	(0.26, 0.1)
v_4	(0.48, 0.42)	[0.48, 0.58]	(0.48, 0.1)
v_5	(0.55, 0.35)	[0.55, 0.65]	**(0.55, 0.1)**
v_6	(0.59, 0.31)	[0.59, 0.69]	**(0.59, 0.1)**

Step 5

We assume that α =0.55, and the reduction is done via consideration of only those pairs, in which the first values are greater than or equal to the value 0.55. Now, after reduction, we consider the following pair of numbers

$$Per_{IFS}(v_5) \prec_L Per_{IFS}(v_6),$$

that is

$$\mathbf{(0.55, 0.1)} \prec_L \mathbf{(0.59, 0.1)}.$$

Step 6

The ordered pairs of the numbers

$$(0.55, 0.1) \text{ and } (0.59, 0.1)$$

are converted, according to Eq. 7.21, to the intuitionistic fuzzy sets

$$\{(v_5, 0.55, 0.35)\} \text{ and } \{(v_6, 0.59, 0.31)\},$$

respectively. They are applied to create the set of elementary selectors, describing the group g_2.

$$R_{g_2}^{0.55} : IF \left((value = v_6),\ 0.59,\ 0.31\right) \vee \left((value = v_5),\ 0.55,\ 0.35\right)$$

$$THEN\ a\ given\ object\ is\ a\ member\ of\ the\ group\ g_2.$$

The generated rule consists of two elementary selectors, one related to the attribute value v_6 and the second to the attribute value v_5. Similar results have been obtained for the multiset representation, considered in Sect. 5.6.4. It is interesting that the classification rules are very similar despite the use of different formalisms.

Table 7.6 The testing intuitionistic fuzzy sets C_{e_n}, $n = 7, 8, 9, 10$

n	v_1	v_2	v_3	v_4	v_5	v_6
7	(0.71, 0.19)	(0.25, 0.65)	(0.66, 0.24)	(0.28, 0.62)	(0.28, 0.62)	(0.24, 0.66)
8	(0.27, 0.63)	(0.65, 0.25)	(0.26, 0.64)	(0.69, 0.21)	(0.69, 0.21)	(0.69, 0.21)
9	(0.71, 0.19)	(0.65, 0.25)	(0.75, 0.15)	(0.28, 0.62)	(0.28, 0.62)	(0.24, 0.66)
10	(0.27, 0.63)	(0.25, 0.65)	(0.26, 0.64)	(0.28, 0.62)	(0.69, 0.21)	(0.69, 0.21)

• *Brief analysis of classification rules*

Let us consider the generated classification rules $R^{0.55}_{g_1}$ and $R^{0.55}_{g_2}$, formulated in the abbreviated form as

$$R^{0.55}_{g_1} : IF \ \big((value = v_3), 0.58, 0.32\big) \vee \big((value = v_1), 0.57, 0.33\big) \ THEN \ g_1$$

$$R^{0.55}_{g_2} : IF \ \big((value = v_6), 0.59, 0.31\big) \vee \big((value = v_5), 0.55, 0.35\big) \ THEN \ g_2.$$

The classification quality of the generated rules $R^{0.55}_{g_1}$ and $R^{0.55}_{g_2}$ was checked by applying the rules to the testing text documents, which did not participate in the generation of these rules, see Table 7.6.

Using Eq. 7.11, we obtained the values of the measures of perturbation (eight intervals) of each test object: e_7, e_8, e_9, e_{10}, for the classification rules $R^{0.55}_{g_1}$ and $R^{0.55}_{g_2}$.

Then, the intervals were converted to the ordered pairs of numbers, using the operation $C_{\rightleftharpoons}(.)$, defined by Eq. 7.20.

The use of the lexicographical ordering allows for selecting of a rule, which matches better the testing object, meaning that the measure of perturbation is lower.

The following measures of the perturbations (intervals) were calculated for the testing examples.

For the object e_7:

$$Per(R^{0.55}_{g_1} \mapsto e_7) = [0.22,\ 0.32],$$

$$Per(R^{0.55}_{g_2} \mapsto e_7) = [0.57,\ 0.67].$$

The lexicographical ordering

$$Per(R^{0.55}_{g_1} \mapsto e_7) \prec_L Per(R^{0.55}_{g_2} \mapsto e_7)$$

shows the lower measure of the perturbation $Per(R^{0.55}_{g_1} \mapsto e_7)$, equivalent to the indication of the better matching of the rule $R^{0.55}_{g_1}$ regarding the object e_7. Thus, the text document e_7 was correctly classified to the group g_1.

For the object e_8:

$$Per(R_{g_1}^{0.55} \mapsto e_8) = [0.57,\ 0.67],$$

$$Per(R_{g_2}^{0.55} \mapsto e_8) = [0.21,\ 0.32].$$

The lexicographical ordering

$$Per(R_{g_2}^{0.55} \mapsto e_8) \prec_L Per(R_{g_1}^{0.55} \mapsto e_8)$$

shows the lower measure of the perturbation $Per(R_{g_2}^{0.55} \mapsto e_8)$, equivalent to the indication of the better matching of the rule $R_{g_2}^{0.55}$ regarding the object e_8. Thus, the text document e_8 was correctly classified to the group g_2.

For the object e_9:

$$Per(R_{g_1}^{0.55} \mapsto e_9) = [0.17,\ 0.27],$$

$$Per(R_{g_2}^{0.55} \mapsto e_9) = [0.57,\ 0.67].$$

The lexicographical ordering

$$Per(R_{g_1}^{0.55} \mapsto e_9) \prec_L Per(R_{g_2}^{0.55} \mapsto e_9)$$

shows the lower measure of the perturbation $Per(R_{g_1}^{0.55} \mapsto e_9)$, equivalent to the indication of the better matching of the rule $R_{g_1}^{0.55}$ regarding the object e_9. Thus, the text document e_9 was correctly classified to the group g_1.

For the object e_{10}:

$$Per(R_{g_1}^{0.55} \mapsto e_{10}) = [0.57,\ 0.67],$$

$$Per(R_{g_2}^{0.55} \mapsto e_{10}) = [0.21,\ 0.32].$$

The lexicographical ordering

$$Per(R_{g_2}^{0.55} \mapsto e_{10}) \prec_L Per(R_{g_1}^{0.55} \mapsto e_{10})$$

shows the lower measure of the perturbation $Per(R_{g_2}^{0.55} \mapsto e_{10})$, which indicates the better matching of the rule $R_{g_2}^{0.55}$ with respect to the object e_{10}. Thus, the text document e_{10} was correctly classified to the group g_2.

Hence, all the considered text documents were correctly classified to the appropriate groups. Thus, the classification rules, whose generation is based on the concept of perturbation of intuitionistic fuzzy sets, fulfill the basic requirements of classification quality.

References

1. K.T. Atanassov, *On Intuitionistic Fuzzy Sets Theory* (Springer, Heidelberg, 2012)
2. K.T. Atanasow, G. Gargov, Interval valued intuitionistic fuzzy sets. Fuzzy Set Syst. **31**, 343–349 (1989)
3. J. Kacprzyk, M. Krawczak, G. Szkatuła, On bilateral matching between fuzzy set. Inf. Sci. **402**, 244–266 (2017)
4. K.T. Atanassov, Intuitionistic fuzzy sets. Fuzzy Sets Syst. **20**, 87–96 (1986)
5. K.T. Atanassov, On intuitionistic fuzzy sets theory, in *Studies in Fuzziness and Soft Computing* (Springer, 1999)
6. E. Szmidt, J. Kacprzyk, Distance between intuitionistic fuzzy sets. Fuzzy Sets Syst. **114**(3), 505–518 (2000)
7. M. Krawczak, G. Szkatuła, On matching of intuitionistic fuzzy sets. Inf. Sci. **517**, 254–274 (2020)
8. E. Szmidt, J. Kacprzyk, Entropy for intuitionistic fuzzy sets. Fuzzy Sets Syst. **18**(3), 467–477 (2001)
9. H.-M. Zhang, L.-Y. Yu, New distance measures between intuitionistic fuzzy sets and interval-valued fuzzy sets. Inf. Sci. **245**, 181–196 (2013)
10. J. Kacprzyk, G. Szkatuła, Inductive learning: a combinatorial optimization, in *Advances in Machine Learning I*. ed. by J. Koronacki, Z.W. Ras, S.T. Wierzchon, J. Kacprzyk (Dedicated to the Memory of Professor Ryszard S. Michalski. Springer, Heidelberg, 2010), pp.75–93
11. M. Krawczak, G. Szkatuła, Bidirectional comparison of multi-attribute qualitative objects. Inf. Sci. **436–437**, 367–387 (2018)
12. P. Intarapaiboon, Text classification using similarity measures on intuitionistic fuzzy sets. ScienceAsia **42**, 52–60 (2016)

Chapter 8
Summary and Perspectives

From the very beginning, Artificial Intelligence dealt with the recognition of system states. Nowadays, the term of Intelligent Computing is used more and more often, and also a separate information granule is being called an object. It is assumed that such objects are described by a set of features, and a specific object is described by a set of values of appropriate features. In general, there are two main issues of operations on such objects, one concerns the classification of objects, the other concerns the grouping of objects. In both of these issues, directly or indirectly, the basic task of comparison of objects is involved. Therefore, assessment of proximity of the compared objects and a proper measure of it, is still a crucial problem, and additionally, comparison of such objects often appears to produce results that are not symmetrical.

The proposed perturbation methodology and related to it directional proximity, allows us to solve many, real problems in which asymmetry of proximity of objects cannot be neglected. The inclusion of directional proximity seems to be appropriate for designing more effective, relatively simple, and intuitive methods of matching of objects, for example, in data mining and Internet mining resources.

Undoubtedly, the most extensive use of classification and grouping of objects is related to Internet data mining. The rapidly increasing resources of information on the Internet necessitate continuous improvement of information retrieval methods. Data mining on the Internet is about querying a specific data pattern. Such a pattern can be called the prototype, and then during the search process throughout the Internet resources some similar objects are found. The objects found can be called variant patterns. It is assumed that the prototype pattern is compared to variant patterns. The process of comparing is done under the tacit assumption that the variant pattern is similar to the prototype pattern, to the same extent, as the prototype pattern is similar to the variant pattern. The result is a list of accepted variants, and each variant is somewhat similar to the prototype. This list of variants, compiled with a specific matching margin, depends on the accuracy of the search procedure. It is obvious that with a larger margin the list of variants grows significantly. However, with such

G. Szkatuła and M. Krawczak, *Bidirectional Comparison of Nominal Sets*,
Studies in Computational Intelligence 1140,
https://doi.org/10.1007/978-3-031-53096-8_8

tacit assumption, in the process of searching for variants, information about possible asymmetry of the compared variants and the pattern is lost.

In order to use the paradigm of asymmetric comparison of objects, let us apply the following procedure. By searching the Internet, comparing the prototype to variants, with a given level of similarity, we get a set of variants. However, it may be interesting to obtain information obtained by comparing the already selected variants to the prototype, i.e. asymmetric information to that used for searching variants. In this way, we obtain additional information about the relationship between the selected variants and the prototype.

Another important issue, considered in the book, are the different ways of describing the nominal-valued attributes, whose values may be repeated, or not exactly known, or incomplete and inconsistent. This book describes attributes using ordinary sets, multisets, fuzzy sets, and also using intuitionistic fuzzy sets. For each type of sets, representing attributes, the concept of perturbation of one set by another set is presented. The concept of set perturbation is the basis of the methodology of directional object comparison, and then, in order to capture the asymmetry of the similarity of sets, the asymmetry of the similarity of the prototypes and variants is considered. In this manner, we are able to obtain additional information when objects are compared, information that may not always be disregarded.

It seems that the next step should be to consider objects represented by mixed information, meaning that each object is described by attributes, and the attributes can be represented by different kinds of sets. It should be emphasized that each individual attribute may be described with the same kind of a set.

Both of the presented forward-looking directions for further research are open problems, and are not trivial, constituting, indeed, a great research challenge.

Index

G. Szkatuła and M. Krawczak, *Bidirectional Comparison of Nominal Sets*, Studies in Computational Intelligence 1140,
https://doi.org/10.1007/978-3-031-53096-8

The manufacturer's authorised representative in the EU is Springer Nature Customer Service Centre GmbH, Europaplatz 3, 69115 Heidelberg, Germany. If you have any concerns regarding our products, please contact ProductSafety@springernature.com

Printed and bound by CPI Group (UK) Ltd, Croydon, CR0 4YY
30/09/2026
02225561-0002